2.30

HOW TO DO A SCIENCE FAIR PROJECT

Revised Edition

HOW TO DO A SCIENCE FAIR PROJECT

Revised Edition

Salvatore Tocci

Experimental Science

Franklin Watts
A Division of Grolier Publishing
New York • London • Hong Kong • Sydney
Danbury, Connecticut

Note to readers: In most cases, measurements are given in both metric or English units. Wherever measurements are given in only one system, the units provided are the most appropriate for that particular situation.

Photographs ©: Bruce C. Stewart/University of Massachusetts at Amherst, College of Engineering: 35; Photo Researchers: 33 (A. Barrington Brown/SS); Randy Matusow: 11, 13, 15, 17, 19, 23, 24, 28, 30, 36, 38, 41, 45, 47, 49, 61, 62, 69, 72, 74, 80, 81, 103, 106, 107, 108, 113, 119.

Library of Congress Cataloging-in-Publication Data

Tocci, Salvatore.
How to do a science fair project / Salvatore Tocci. — Rev. ed.
p. cm. — (Experimental science)
Includes bibliographical references and index.
Summary: A step-by-step guide for creating a variety of projects suitable for entry in a science fair with suggestions for choosing a subject, performing the experiment, and polishing the presentation.
ISBN 0-531-11346-9 (lib. bdg.) 0-531-15881-0 (pbk.)
1. Science—Experiments—Juvenile literature. 2. Science—Exhibitions—Juvenile literature. [1. Science—Experiments. 2. Science—Exhibitions. 3. Experiments.]
I. Title. II. Series: Experimental science series book.
Q164.T681997
507'.8—dc21 96-50019
CIP
AC

Printed in the United States of America.
1 2 3 4 5 6 7 8 9 10 R 06 05 04 03 02 01 00 99 98 97

CONTENTS

INTRODUCTION

What Is a Science Fair Project?

Doing a science fair project is a good way to learn about science and how scientists think and work. You may think that science consists mainly of factual information contained in some book and that scientists always make their discoveries by following a set routine. You may even believe that science is primarily limited to schools, libraries, and research laboratories. But science is more than a subject taught in school, the topic of books in libraries, or a procedure followed in a laboratory. Science is a search for answers.

One of the best ways for you to become involved in such a search is to conduct a science fair project. This book will show you how to carry out a project and tell you how scientists approach problems in searching for an answer or a discovery. To get started, all you need is a good idea and some curiosity. The only limits on what you can accomplish might be those of your imagination.

This book will discuss each step of the process—from selecting a topic to presenting your project at a science fair. Although it covers the kinds of projects usually done for a science fair, the book emphasizes projects based on original investigations. To complete this type of project, you must conduct experiments in an effort to answer a specific question or solve a particular problem. Chapters 4 and 5 contain information and suggestions only for projects involving original investigations. These projects require more thought and effort than those limited to collecting information from encyclopedias or building a model from a store-bought kit.

No matter what kind of project you choose, what you learn will make you more knowledgeable about the world and how it works. Your project may even win an award. But beware—although you may heed all the advice and suggestions offered in this book, you may still face disappointments, get trapped in blind alleys, or follow misleading clues. Don't get discouraged; that's what a science fair project is all about—the real world.

Unfortunately, some science fair projects look like class assignments rather than the tangible output of a fertile imagination and a serious investigation into the real world. You may be planning to do a project because it is a required part of your science course. A certain percentage of your grade may even depend on your project. With little time and practically no preparation, you might resort to constructing a poster display of pictures cut from magazines, writing a report from an encyclopedia, or performing a lab exercise obtained from a textbook. Such projects are not real science but only "cookbook" chemistry—just follow the recipe, and the result is guaranteed.

So, even if your science fair project is a required assignment, try to approach it with enthusiasm and a desire to discover something about the world. Follow the suggestions in this book to make the assignment a learning experience. Because science involves challenges to be faced and solutions to be discovered, a science fair project should reflect a sense of excitement as well as creativity, curiosity, and achievement.

If you think about these qualities, you'll see why a scientist can be considered a detective. Both are looking for an answer, with only a few ideas or clues leading the way to the final solution. By conducting a science project that seeks to answer a question raised by your curiosity, you'll be carrying out the same mental processes used in detective work.

To see the connection between conducting a science fair project and a criminal investigation, you may want to read about one of the world's greatest detectives, Sherlock Holmes. Holmes's investigations were always challenges in which the solutions were found by using a logical, scientific process, much like the one needed to carry out a science fair project. Before undertaking your project, you might find it worthwhile to read about some of Holmes's cases in *The Complete Sherlock Holmes*.

CHAPTER 1

Selecting a Topic

Once you've decided to do a science fair project, you need to select a topic. If you look no further than your science textbook for ideas, choosing a topic can be difficult. In fact, selecting a topic can be the most challenging part of your project, especially if you don't know where to begin.

Start by imitating Holmes—depend upon your ability to observe. When Holmes meets Dr. Watson for the first time, he tells him, "Observation with me is second nature." Holmes proves his point by telling Watson what he knows about him simply by observing his physical appearance: Watson is a doctor who has undergone recent hardship and sickness while serving with the British army in Afghanistan. If Holmes can get so much information from a quick observation, think what you can learn from a more thorough study.

WHERE TO OBSERVE

Start by observing the world around you, and you might discover an idea for your project. You can visit a museum, look under the kitchen sink, walk through a forest, watch a movie, or go to a zoo. Look closely and be alert for the unexpected. Ideas for a project can arise without warning. Many important discoveries have been made because scientists have taken notice of something totally unexpected. Perhaps the best-known unexpected finding in the history of science was the discovery of penicillin.

Visiting a museum such as the National Air and Space Museum in Washington, D.C., may help you think of a science fair project idea.

In 1928, Scottish bacteriologist Alexander Fleming was working with bacterial cultures. Absorbed by his project, Fleming didn't bother to clean the culture dishes after he had finished with them. Mold soon started to form in these dishes. Even though his attention was focused on his work, Fleming observed that the areas surrounding the mold in these dishes were free of bacterial growth.

Fleming thought the mold might contain some sort of chemical substance that inhibited the growth and reproduction of bacteria. He abandoned his original project to investigate his new idea. Fleming confirmed his suspicion when he placed samples of the mold in different media and found no bacterial growth. Fleming's keen observation and interest in pursuing an unexpected finding led to the discovery of penicillin, an antibiotic that was to save thousands of lives during World War II—and millions thereafter.

OBSERVE CLOSELY AND CAREFULLY

Don't underestimate the importance of making observations. Observation plays a fundamental role in science, as well as in detective work. Many young scientists, however, fail to recognize that observation can be an extremely complex activity and can reveal a great deal of information. If you want to test your powers of observation, try the following task. You will need a candle, a match, and a ruler. Describe the candle as completely as possible; then light it and make additional observations.

Do not confuse observations with *inferences*. You are making an observation if you say that a colorless liquid collects near the wick. But if you say this liquid is melted wax, you are making an inference. An inference is a conclusion based on indications or logical reasoning and not on direct proof. An observation involves gathering information using any of your five senses. See how many observations you can make. Observe carefully, because you may overlook something if you don't pay close attention.

PURSUE YOUR INTERESTS

Focus your observations on something that genuinely interests you. When Dr. Watson asked whether any cases were open for investigation, Holmes replied, "Some ten or twelve, but none which present any feature of interest. They are important, you understand, without being interesting." Do not select a topic on the basis of its importance to others but rather on its interest and appeal to *you*. After all, you will be spending a considerable amount of time on the project, so you might as well enjoy what you are doing.

The most important factor to keep in mind when choosing a topic is to pick one that is based on your interests. This student may want to do a project that examines how various factors influence model race car performance.

Look at your favorite hobby for a science project idea. Perhaps your pet or some sport might inspire you. If you have an aquarium at home, you could design and construct a more efficient filtering system, or study fish behavior under different lighting conditions. If you play basketball, you could test various sneakers for their durability, comfort, and exactly how high each brand enables a person to jump. In any case, consider your personal interests.

OTHER SOURCES OF IDEAS

If your hobbies fail to provide any ideas for a project, don't give up—there are many other places to search. Sources for ideas include books, magazines, encyclopedias, on-line computer services, special publications, television shows, and people.

Glance through your science textbook to review the topics you covered in class. Did you find a particular chapter interesting or do exceptionally well in one area of the course? Reread those sections in the book to refresh your memory and perhaps provide the spark for a creative idea. While reading the chapter, remember you are looking for ideas. Pay particular attention to topics in which research is currently being conducted or where problems remain to be solved. Refer to the end of that chapter for independent investigation ideas or suggestions for further reading.

Another source of ideas is a library, either the one in your school or your community. Check the catalog holdings for titles of science books, but don't be limited to titles that sound like textbooks. Check out those with provocative titles like *Psychic Phenomena* or *Creatures Beneath the Seas*. These books may deal more with fancy than fact, but they could provide you with an idea. For example, many scientists are trying to establish a scientific basis for extrasensory perception (ESP). Others are looking for deep-sea organisms that may be the ancestors of modern fish. If you head in this direction, make sure your idea for a project is scientifically sound and does not fall into the area of pseudoscience.

If you have no success with books, ask your librarian for assistance in locating magazines that can provide ideas. Possibilities include *Scientific American*, *Science News*, *Discover*, *Smithsonian*, *Natural History*, *Omni*, and *Popular Science*. Numerous scientific journals and publications are also available. They provide current information on a wide variety of topics.

Your library will probably not carry all these publications. Check with your librarian for CD-ROM programs that contain data banks of articles from a wide selection of periodicals. In many cases, you can print out the entire text of an article. If

This student is searching for information on a CD-ROM. Many libraries have CD-ROMs that provide the complete text of recent periodical articles published in a particular subject area.

only a short summary is available, you can ask your librarian to try to locate the original article for the full text.

Also check the *Reader's Guide to Periodical Literature*, an index containing lists of articles published in many magazines. Listing the author's last name, title of the article, and a short synopsis, this guide will direct you to the specific volume and pages of the magazine where the article appeared. In some cases, your school library may not have the magazine on file; a trip to a college library may be necessary.

Some of these professional scientific journals may be too advanced for you to understand the material. Try to scan these magazines to see if you can pick up some project ideas without worrying about all the details. Or you may find some publications too "sensational." Look at the "science" in the article and don't be distracted by the sensationalism. Finally, look for articles whose findings are controversial. When scientists are still debating the findings of a particular study, your project may shed some additional light on the subject.

For example, a recent science magazine reported that marathon runners and other endurance athletes appear to perform better after consuming a high-fat diet for several weeks than after eating their usual low-fat, high-carbohydrate diet. Three groups of athletes were chosen for the study. All three groups were fed diets containing the same number of calories. But one group got 16 percent of their calories from fat; the second group, 30 percent (the recommended amount for most people today); and the third group, 45 percent. After completing a 4-week diet, the athletes on the fattier diets were able to increase the amount of time they could run at peak capacity.

Some scientists question the findings of this report, however, pointing to the health risks associated with high-fat diets. What do you think? After reading about such studies, you might want to talk with your science teacher. He or she can help you develop your idea. You may decide to design an experiment using small rodents to test whether a high-fat diet enables these animals to increase their stamina when exercising.

Your science teacher can help you develop your idea and may even know someone who works in the area you have decided to explore.

ADDITIONAL LIBRARY SOURCES

Check your library for *Facts on File* and *SirS* (*Social Issues Resources Series*). *Facts on File* contains numerous suggestions for experiments, demonstrations, and projects. Each suggestion comes with a list of required materials, a detailed procedure to follow, data sheets to fill in as you work, and an estimate of how long the exercise will take. Each volume of *SirS* includes articles chosen from a variety of science publications. The articles are listed chronologically and a cross-reference guide to articles in all the volumes is provided.

Undoubtedly your library will have several encyclopedias. The most popular scientific encyclopedia, according to a recent study, is Grolier's *New Book of Popular Science*, which provides information about the major fields of science and describes their applications in the world today. Each volume has a bibliography that lists and briefly evaluates a number of books on related subjects. Articles in *Encyclopaedia Britannica* are detailed and written by specialists in their fields.

Encyclopedia Americana is a good source for articles that discuss various aspects of technology. *World Book Encyclopedia* actually lists project ideas in some of the science articles. *McGraw-Hill Encyclopedia of Science & Technology* specializes in science, mathematics, and engineering and is more technical than the other encyclopedias.

When checking reference materials, be sure to scan the multimedia encyclopedias that are available through on-line computer services such as America On-Line, Prodigy, CompuServe, and Microsoft Network. You can also use a local on-line access company that provides service to your area. These companies generally provide a local telephone number to dial when accessing their services through a modem. In addition to encyclopedic information, most of these services have other useful reference resources. Moreover, these services provide access to the Internet.

THE WORLD WIDE WEB

The Internet consists of a worldwide collection of computer networks, with each network in turn consisting of a network of computers operated by government agencies, businesses, universities, research laboratories, and even homes. In effect, the Internet is the world's biggest "information mall." You can "shop" or search for anything, including ideas for your science fair project.

The part of the Internet that many people find useful in searching for information is the World Wide Web. Besides using one of the on-line services mentioned previously, you

These students are using computers to look for information about the topics they have chosen. Ask your school and local librarians for assistance in using multimedia encyclopedias and on-line services.

can reach the Internet through specialized computer services known as Internet service providers or ISPs. These companies feature access only to the Internet and the Web, and are generally less expensive if you use the Internet for more than 30 hours per month.

No matter how you get there, searching, or "surfing" the Web can be somewhat intimidating. Using a directory, which consists of hundreds of thousands of documents sorted into various categories, can make your search easier. Directories are most helpful for general searches. For example, if you are interested in doing a project on diets but don't know exactly what you would like to do, check for a directory or category labeled "Health." The most popular directories are Yahoo and Magellan.

If you go to the Yahoo home page at http://www.yahoo.com/, you'll see that the page is arranged by broad areas such as *Business and Economy*, *Education*, and *Health*. From here, you can narrow your search. Unlike Yahoo, Magellan reviews Web sites and lets users know which ones it feels are particularly helpful. The address for Magellan is http://www.mckinley.com/.

For a more focused search, you can use an index with the help of a tool known as a "search engine." Several search engines and their addresses are listed in Table 1. Several indexes and directories on the Internet are free. To use an index, all you do is type in one or more keywords. For example, if you want to do a project that involves professional sports, type in the keywords "professional sports." In this case, the index may uncover too much information. Sorting through it all would take too much time.

Therefore, try to be as specific as possible when using an index. Suppose you want to examine whether performance is related to the salary a player receives. Type in the keywords "salaries in professional sports" to zero in on what you are looking for without having to sort through long lists.

TABLE 1 ACCESSING AN INDEX ON THE WEB

Search Engine	*Address*
AltaVista	http://altavista.digital.com
Excite	http://www.excite.com/
Inktomi HotBot	http://www.hotbot.com/
Infoseek	http://guide.infoseek.com/
Lycos	http://lycos.cs.cmu.edu/
Open Text	http://www.opentext.com/
WWW Worm	http://guano.cs.colorado.edu/wwww/
Webcrawler	http://webcrawler.com

Each search engine has its own search techniques. Some will require that you put quotation marks around the words to show that they go together. Others may require that you put the word "and" between the two words so it knows you're looking for a site that contains both words instead of either one of them. Check the search engine's instructions to find out the most efficient way to conduct a search.

CHECK PAST SCIENCE FAIR PROJECTS

Check publications and pamphlets describing past science fair projects for project ideas. Each year Science Service (1719 N Street, N.W., Washington, D.C., 20036) publishes the *abstracts* from the International Science and Engineering Fair (ISEF). The abstracts are short descriptions of the projects entered in the fair that year.

Science Service also has information on the Science Talent Search for Westinghouse Science Scholarships and Awards. A winning project in this competition can lead to a scholarship that will cover 4 years of college expenses! Abstracts of previous projects may also be available from local, regional, and state science fairs.

Although reading about past projects may sound boring, it is just the opposite. *Science Service Abstracts* in particular is a gold mine. If this publication doesn't lead you to several ideas, it will at least show you how ingenious other students have been. This is bound to propel you to new and greater heights!

Don't hesitate to build upon the work of others. Many scientific achievements have their roots in the past. For example, vaccination against diseases such as measles and polio was not possible until a scientist developed a method for growing viruses in glass flasks. Once this technique was perfected, scientists could explore the development of vaccines against viral diseases.

Another scientific discovery with its roots in the past is the process by which sperm and egg cells are produced. Once scientists understood how nonsex cells divide, they could predict the process by which sperm and egg cells were formed—even before they actually observed it! Obviously, previous scientific investigations had laid the foundation for new discoveries.

EDUCATIONAL TELEVISION SHOWS

How many times have your parents told you to turn off the television and do your homework? Here's your chance to do your homework by watching television! Check a newspaper or guide for a listing of upcoming telecasts. Look for shows such as *NOVA*, *National Geographic* specials, *Hidden Worlds*, or any scientific special produced for the Public Broadcasting System. If you have cablevision or satellite reception, check channels featuring programs that examine scientific explorations and findings. The Discovery Channel and the Learning Channel are good places to start.

Many of these shows feature recent advances in scientific knowledge. They often interview scientists who are carrying out projects in medicine, engineering, space exploration, animal behavior, oceanography, computer technology, and genetics. The producers of the show sometimes offer a script and additional resource materials, including background information and questions needing further study.

TALK WITH SOMEONE

If you still aren't sure what you want to do for a project, you may want to talk with professionals in your community. These people work in hospitals, universities, government agencies, environmental organizations, zoos, museums, industries, computer centers, pharmaceutical companies, greenhouses, observatories, botanical gardens, pharmacies, and water-treatment plants. All of these places should be listed in your local telephone directory.

Professionals who work at these places are usually eager to help young scientists and may give special assistance if your project is in their area of expertise. But their time may be limited, so call or write a few weeks in advance to ask for an

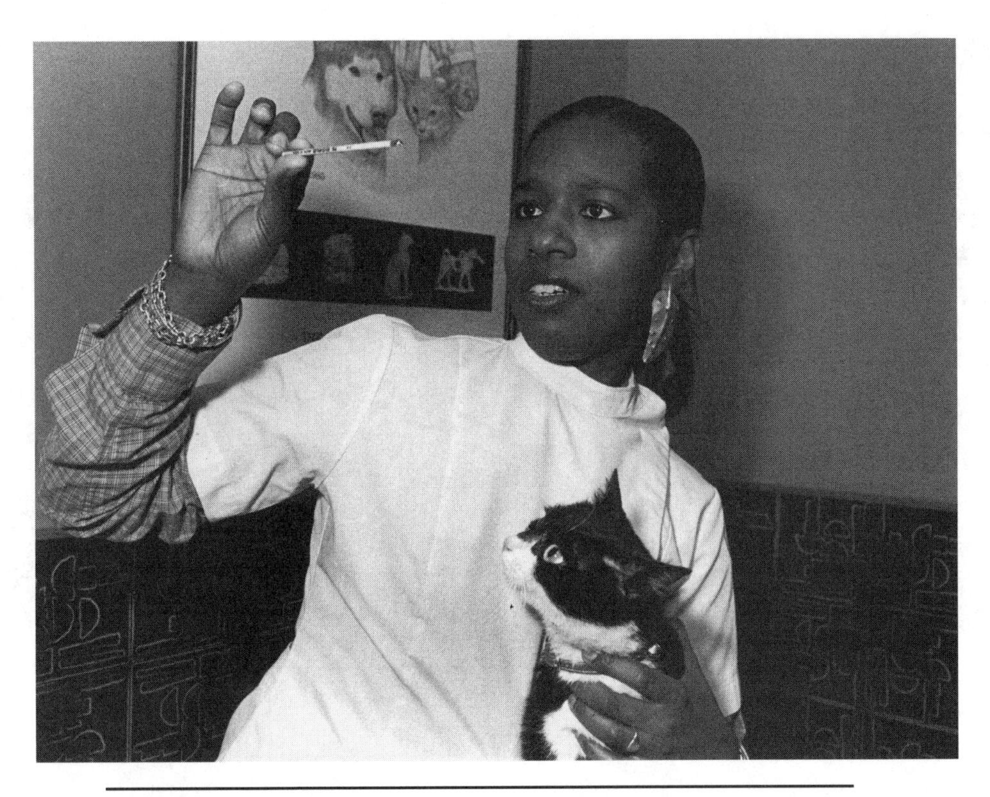

A local professional, such as a doctor, computer scientist, or this veterinarian, may be able to help you develop an idea for a science fair project.

appointment. State what information you want to obtain and request copies of any written materials available on the topic. Before the interview, prepare your questions; during the interview, take notes or tape-record the conversation—after obtaining permission.

If you can't meet with a professional, perhaps a teacher can arrange for a brainstorming session. In such a session, you and other interested students can explore ideas and suggestions for projects. The group may find it helpful to begin by talking about mutual interests, possibly starting with courses you enjoyed or field trips you found interesting.

You and your friends may want to get together in the library to explore ideas and discuss what you find.

FEATURES OF A GOOD TOPIC

Make sure your project idea is realistic and can be accomplished with available resources. You may be interested in atomic structure (a topic on which many books have been written), but you will not be able to probe certain mysteries of the atom without a particle accelerator, which can be found in only a few laboratories around the world.

Your selection of a topic will be limited by your access to the equipment and materials required to conduct your project. In addition, you may require live specimens, chemicals, measuring devices, and special lab apparatus. Check your school science laboratory to see what equipment and supplies are available. A nearby research facility or industry may have specialized equipment not found in schools. For example, you may be able to get access to an X-ray machine to irradiate specimens, a high-speed centrifuge to create ultragravitational forces, or a powerful telescope to observe the stars.

If the project can be accomplished with available resources, remember that *you* will have to do it! You may occasionally require advice and suggestions from a teacher or professional scientist, but the project must reflect your own ideas and efforts. Be sure you can wrestle with the topic by yourself, with only occasional support from others. If you plan to submit your project for evaluation in a science fair competition, you will be judged on *your* knowledge and input into the investigation. Little credit for initiative and creativity can be given to someone whose project reflects the work and ideas of others.

Finally, be sure that you will have enough time to complete the project. If you select a topic that is too broad or generalized, you will not only have difficulty planning the project, but you will also lack the time to arrive at any conclusions or results.

After choosing your topic, narrow it down to a specific question or problem that can be solved within a reasonable time. Too often, students end up with poor project results because they failed to focus on a specific question or problem.

CAN YOU IDENTIFY A GOOD TOPIC WHEN YOU SEE ONE?

You might find it helpful to examine the following list to determine which topics would make a suitable science fair project. Why are some topics inappropriate for a project? Can you suggest improvements or changes that would make these titles more likely to lead to a successful science fair project?

Is There Life on Mars?
Working with Plants
Effects of Radiation on the Growth of Geraniums

The World of Insects
Regeneration of Limbs in Salamanders
Determination of the Radii of Subatomic Particles
Measuring the Efficiency of Solar Collectors
Courtship Behaviors of African Wildlife
Methods of Web Construction in House Spiders
A Seashell Collection from Three Caribbean Islands
Principles of Aerodynamics
A Method to Increase the Fruit Yield of Banana Trees
Recombinant DNA Technology
Estimating Chemical Bond Angles from Nuclear Magnetic Resonance
Which Antacid Is Most Effective?
The Effects of Acid Rain
A Motion Detector-Based Light Switch

Some titles are too vague: "Working with Plants," "The World of Insects," "Principles of Aerodynamics," "Recombinant DNA Technology," and "The Effects of Acid Rain." Left as they are, these titles do not provide a good idea of where to begin. For example, you would have a better idea of where to start if you changed "Working with Plants" to "How the Moisture Content of Soil Affects the Growth of Tomatoes." The latter is far more specific.

Other titles might require resources that are not easy to obtain: "Is There Life on Mars?" and "Courtship Behaviors of African Wildlife." Such projects would require unlimited funds, access to highly sophisticated equipment, or months of free time.

Two of the suggested titles might be beyond the capabilities of a young scientist: "Determination of the Radii of Subatomic Particles" and "Estimating Chemical Bond Angles from Nuclear Magnetic Resonance." These projects require highly sophisticated equipment and technical knowledge. And, in the end, such projects may not reflect *your* thinking and effort but rather those of a professional scientist.

The remaining titles represent better ideas because they are more specific, realistic, and capable of being accomplished within a reasonable time. If you picked the last title, you made a good choice—this is the title of an award-winning project. A brief look at this project might help you see what steps to follow in selecting a good topic.

AN AWARD-WINNING IDEA

Stanley Lin developed an award-winning project as a result of the steps he followed in choosing his topic. As a high school sophomore, Stanley became interested in electronics and decided to take a course called Science Research Projects.

Stanley wanted to build a remote-controlled robot that would serve as a "butler." The robot would have an infrared sensor that would respond to signals in a control unit that Stanley would design and build. But, unfortunately, Stanley encountered several problems. He soon recognized that he had selected a project that he could not complete within the alloted time. He also had trouble locating reference materials. Although Stanley did not accomplish what he had set out to do, he used what he had learned from his experiences when he undertook a project the following year.

Stanley knew that he wanted to do an electronics project that would have a practical application—something that people might find useful in their daily lives. After speaking with his teacher, he sought additional information by reading some materials he obtained at a local college library and talking with a friend who had a similar interest in electronics.

Notice the steps Stanley took to choose a topic for his project in his junior year: (1) looking at a hobby for ideas, (2) talking with a teacher, (3) reading library materials, (4) talking with a friend, and (5) learning from his experiences with a prior project. Only then could he decide what he was going to do—build a motion detector that would turn on the lights when someone entered the room and then turn them off when

If you follow the same steps as Stanley Lin in developing a project, you too may be a winner. The inset shows the award-winning motion-detector apparatus that Stanley designed and built.

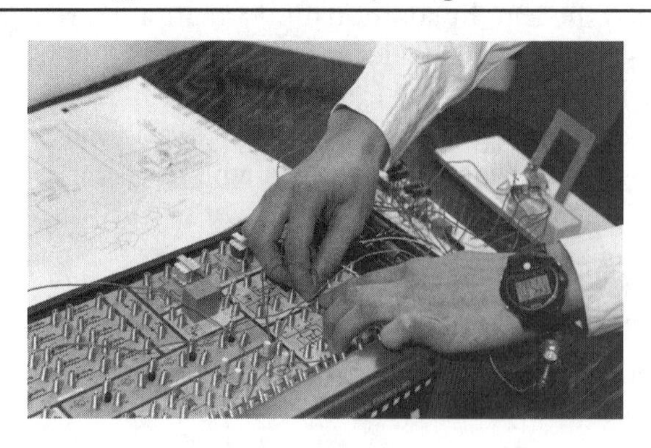

the person left. Once he was sure that he could obtain the necessary equipment and materials, Stanley began his project. This time he was able to complete his project in time for the county-wide science fair where he won an award.

By the way, if you think you might do another project at some future date, choose one that you can expand as you proceed. For example, Stanley's project operates whether it is day or night. Obviously, a motion detector light switch has little, if any, value during the day. Consequently, Stanley is considering modifying his motion detector by incorporating a photosensitive relay that shuts off the unit whenever a certain amount of light is present.

WARNING: SOME TOPICS REQUIRE SPECIAL CONSIDERATIONS

Most science fairs have rules and regulations concerning certain projects. In some cases, you may have to modify or change your topic because of these rules. To save yourself a lot of unnecessary work, find out the rules even before you start looking for ideas. The following three rules are part of most science fair regulations.

First, any *vertebrates* (animals with backbones, such as fish, frogs, birds, and mammals) used in a project must be given every humane consideration for their comfort and treatment. Whenever possible, use plants, single-celled organisms, or *invertebrates* (animals without backbones, such as worms, insects, crabs, and jellyfish). In fact, invertebrates make excellent test subjects. However, if your project requires the use of vertebrates, you must have adequate knowledge about their behavior, characteristics, needs, and handling.

Proof of qualified adult supervision from a teacher or research scientist may be required by some science fair competitions before accepting a project involving vertebrates. This is what Jason Williams did when he carried out a project investigating the effects of light on mice.

If you decide to work with a vertebrate, be sure to do what Jason Williams did—complete the necessary permission forms and have an adult supervise your project.

Any project involving inhumane treatment cannot be included in a science fair. Such projects raise moral and ethical questions. You must also follow recommended methods for sacrificing animals if this is required to conduct your project. Contact Science Service for additional information on projects involving vertebrates.

Second, check with Science Service if your project will involve human subjects. Some science fairs prohibit the use of human subjects. Again, supervision by a qualified person is required for such a project. These guidelines apply even if you plan to be the subject of your own project.

Third, a project involving genetic engineering must be carried out within well-defined guidelines. Genetic engineering is a recent scientific breakthrough in which an organism's genes, which are composed of a chemical substance called DNA, can be altered or rearranged. If your project involves genetic engineering, only research that is normally conducted in a microbiological laboratory with special containment facilities is permitted. The supervision of a qualified scientist is also required. Write to the National Institutes of Health (9000 Rockville Pike, Bethesda, Maryland 20892) for information on experiments involving genetic engineering.

As Holmes tells Dr. Watson, "You did not know where to look, and so you missed all that was important." You now know where to look for ideas and recognize important factors to consider when selecting a topic.

CHAPTER 2

Types of Projects

Once you have selected a topic, your next step is to examine the various ways you can conduct your project. The type of project you decide to undertake will affect how you should proceed with your work. Nearly all science fair projects fall into one of five categories: assembling a model, constructing a display, carrying out a survey, repeating someone else's work, or undertaking an original investigation.

MODELS IN SCIENCE

Models have played a major role in some important scientific discoveries. Scientists have used pieces of wood, plastic, and metal to construct models of atoms, chemical compounds, and cellular structures. In some cases, these

Models, such as this model of DNA, have played an important role in science. Rather than use tangible objects, you may want to use a computer program to design and construct your model.

models were so successful that the scientists received a Nobel Prize, the highest level of recognition in their field.

Perhaps the best-known example is the discovery of the structure of the chemical compound deoxyribonucleic acid, known simply as DNA. Based on previous experiments, scientists knew that DNA is responsible for hereditary characteristics such as the color of our eyes or the curliness of our hair. Although they had identified all the components in DNA, scientists did not know how all these pieces were arranged. Three scientists, James Watson, Francis Crick, and Maurice Wilkins, attacked this problem by building models. They finally succeeded in building a model that correctly showed the arrangement of all the pieces. In 1962, they received a Nobel Prize for their work.

Models can also be a creative outlet for one's imagination. Consider the work of the famous engineer and architect Buckminster Fuller. In the 1920s, Fuller began using special metals known as alloys to build unconventional structures. He experimented with various models before constructing his first geodesic dome, a structure consisting of triangular parts made of lightweight metal. Fuller designed more than 2,000 geodesic domes. Today, these structures can be found in more than 30 countries.

Scientists are constantly building models to test for more efficient designs of cars, planes, and other machines. They often use computers to create models, not only of the submicroscopic world but also of the cosmos. On the cosmic level, computer models are used to predict natural catastrophes, including earthquakes and hurricanes, and to study trends in global temperature changes over the years. In creating such models, scientists often input data gathered from a variety of sources, including satellites, reconnaissance aircraft, and onsite recording devices.

MODELS IN SCIENCE FAIRS

Models are the format of choice for many elementary and junior high school students. With easy-to-use materials such as wood, metal, and cardboard, you can quickly and easily construct a model of the solar system, the human eye, a frog's internal organs, or a volcano.

Such models may be detailed and colorful, but building them rarely draws upon your thinking, reasoning, or creative skills. In fact, models based on textbook drawings or built from kits are not allowed in some science fairs. These models are

These engineering students at the University of Massachusetts at Amherst are attempting to build the world's most fuel-efficient vehicle powered by an internal combustion engine. They will enter it in an annual competition sponsored by Briggs and Straton, Eaton, and the Society of Automotive Engineers.

excluded because they do not reflect any originality or an attempt to explore a project the way a scientist would.

For such science fairs, an acceptable model must include some original thinking, present a creative solution to an old problem, or constitute a more efficient system. Adding flashing lights or colorful labels reflects little originality. Thousands of projects have featured models of volcanoes complete with "fire" and flowing "lava." Year after year, such models fail to impress the judges at science fair competitions. If you decide to construct a model, build one that displays creativity or has a practical application.

For his project, Mazdak Hobbi built a solar-powered model airplane. The idea for his project came from two sources: an interest in solar power and a television show he saw on model planes. Mazdak checked numerous reference sources, including the Web. He uncovered many references to solar-powered devices, but not a single one to a solar-powered model airplane. Thus his project would certainly be original and creative.

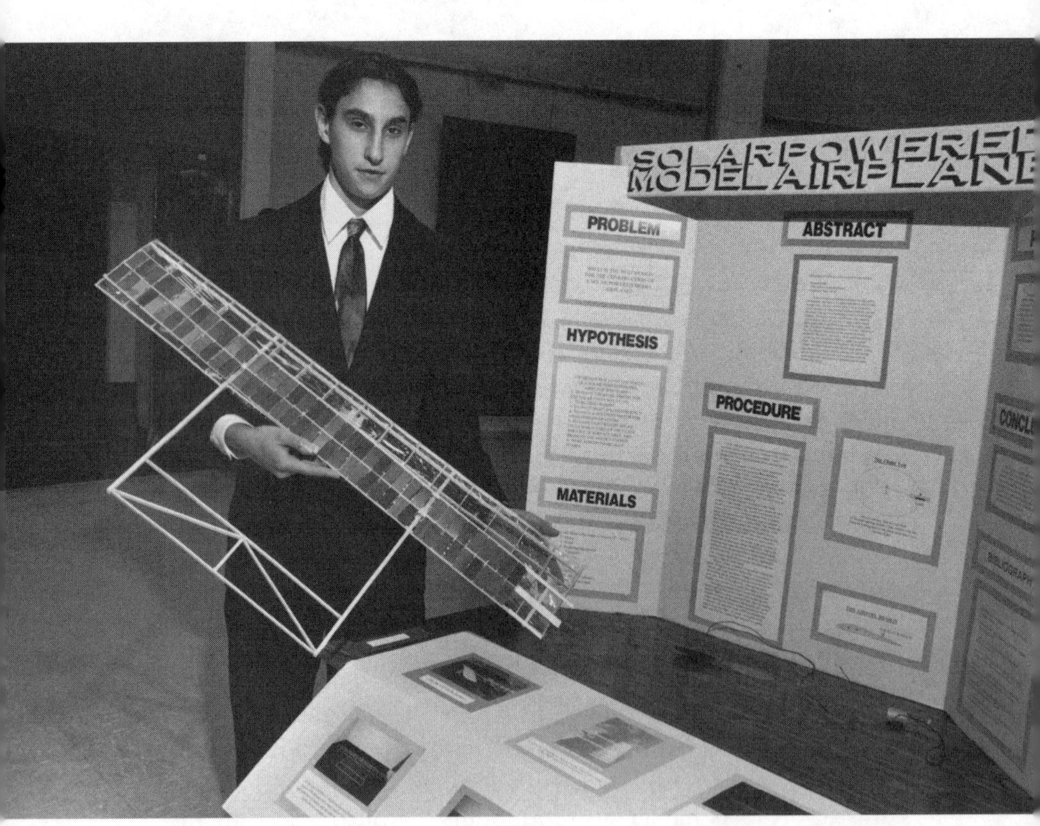

Mazdak Hobbi used balsa wood, plastic wrap, and glue to build this model of a solar-powered model airplane.

Before beginning to build his model, Mazdak talked on several occasions with an aeronautical engineer who taught him the basic principles of aerodynamics. Mazdak then set about building prototypes out of cardboard. In all, he designed and built six prototypes, including two that he felt were viable candidates for building his solar-powered unit.

Mazdak's model is built entirely from balsa wood, plastic wrap, and glue. The power pack consists of some 80 solar cells that can provide a maximum of 10 volts to propel the plane. When released at standing height, the plane can climb 1 meter (3.3 ft.) per second to reach a height of approximately 3 meters (10 ft.). The maximum time it has remained aloft is 7 seconds. The plane has been constructed so that it flies in a controlled circular path before it glides back to the ground.

Mazdak plans to improve his award-winning project for his next competition. He intends to include a remote-control unit that will direct the flight path with the help of an on-board microprocessor. The microprocessor will also control a microcamera that transmits video signals to a receiver on the ground. Mazdak feels that his solar-powered plane might be cost-efficient for surveyors and others who need to obtain aerial views of the topography of a particular area. But regardless of all the technology Mazdak will incorporate into his new model, one component will harken back to the days of the Wright brothers. Without a rudder, Mazdak's model will turn using the same principle the Wright brothers used: the wing will tip down to make the plane turn in that direction.

Models can also serve as prototypes for the "real" thing. With such projects, a team effort may be needed. Cristy Bonuso and Neetu Dhadwal titled their project "The Creation of a Motor- and Crank-Driven Adjustable Device for Walking." Neetu got the idea for the project from her experiences as a volunteer hospital worker. Part of her responsibilities included helping elderly people with their physical rehabilitation program. This work often involved assisting people with their walkers as they tried to regain strength and mobility in their legs and arms. Neetu noticed that the walkers, although helpful, often restricted where the patients could go and certainly prevented them from using staircases during their walks.

Neetu spoke with Cristy, and the two set about designing a more flexible walker that could be used on stairs. They sought the advice of a professor from a nearby university and under his guidance, constructed and tested several models. Their final model was built entirely from PVC piping and joints that were glued with epoxy. The two devices visible on either side of the walker are cordless screwdrivers.

When activated, each screwdriver turns a nut-and-bolt assembly that either raises or lowers that side of the walker, depending on whether the forward or reverse gear is used. In

Cristy Bonuso and Neetu Dhadwal collaborated to design and build this motor-driven walker that would allow disabled people to walk in places that are inaccessible with an ordinary walker.

addition, each side has a manual crank assembly that allows the operator to bring the front and rear bars closer together. In that way, the walker can be made narrow enough for use in going up and down stairs.

Although they received one of the highest awards at a regional science fair competition, Cristy and Neetu are not satisfied with their model. They feel the weight, which is more than that of a conventional walker, is a drawback to their design. They are planning to replace the screwdrivers with lighter-weight motors that can operate the nut-and-bolt assemblies. Perhaps their prototypical model will someday replace the walkers currently in use.

By the way, notice that projects, such as the one done by Cristy and Neetu, can represent the combined efforts of two or more students. If you are thinking of working with someone else on a project for a local science fair, be sure to get permission from your teacher first. If you plan to enter a regional or state competition, be sure to check for any regulations regarding projects carried out by more than one person. Keep in mind that people will expect more from a project done by two or more people than from a project carried out by only one person.

DISPLAYS

Another type of project is a display that consists of a combination of specimens, photographs, drawings, and written reports. A display can be attractive and contain a good deal of scientific information.

Suppose you decide to conduct a project tracing the history of flight. Pictures of the first flight by the Wright brothers, photographs of various jet aircraft, and written information on the space shuttle can be arranged into an attractive display. The visual and written information displayed in your project could provide an impressive summary of the history of aviation.

A display can also reflect a hobby or some leisure time activity. For example, an interesting arrangement of rocks, butterflies, or leaves would make an attractive display. By now, however, you should realize that, like a model, a display may be quite elaborate but still unoriginal. Whether the display represents a hobby or a collection gathered on field trips, adding some scientific information about the items would be valuable.

Perhaps you can show some originality by trying to discover several interesting scientific facts about your collection. Rather than showing all your butterflies, try selecting any that have striking resemblances. Even though two butterflies have been placed in separate categories of classification, they may be similar in coloration, wing structure, or body shape. Your project can explore the scientific reason why these specimens were placed in separate classification groups. Perhaps you can extend your project into an investigation of living butterflies, both in nature and under controlled laboratory conditions. You may arrive at some interesting observations that can then become part of your project.

A project that includes leaves exposed to different environmental conditions would be more interesting than a collection of leaves from the trees in your backyard. Check with local environmental groups to identify any threat to your community, such as acid rain, air pollution, or chemical-waste disposal. These problems are quite serious in some areas of the country; a scientific examination of the local foliage may reveal some hidden impact on your community. In any case, a project involving a scientific look into the materials on display would be more impressive than simply mounting and identifying the items.

This is exactly what Gaetana Barba did. Titled simply "Automobile Safety," Gaetana's project involved a display of the various safety features incorporated into today's cars. She displayed several photographs of cars equipped with safety belts, air bags, and child-restraining seats. Information about each type of safety device was also posted, including data showing the number of lives saved and serious injuries prevented with the proper use of each type of safety device.

Gaetana Barba's display of automobile safety features included a working model.

But Gaetana added a creative touch to her display. She decided to incorporate a model that would test the relative effectiveness of each safety device. Gaetana obtained a wooden track car from her science teacher and modified it for her project. A balloon was attached to act as an air bag, rubber bands were used as safety belts, and rubber stoppers were added to serve as shock absorbers. A raw egg represented the driver.

With a system of pulleys and weights, Gaetana could start the "car" and have it travel at various speeds along a flat surface. She used a photogate timer connected to a computer to determine the vehicle's simulated acceleration and rate of speed. The effectiveness of each safety device was determined by examining the condition of the egg after the "car" had smashed into a block of wood at various speeds.

Unfortunately, Gaetana discovered that her criterion for determining the effectiveness of each safety device—the condition of the egg after impact—was not as informative as she had hoped. Consequently, she had difficulties in assessing the "extent of damage" and therefore could not arrive at any definitive conclusions. However, her teacher recognized Gaetana's creative contribution to an otherwise straightforward display and encouraged her to enter her project in a regional competition.

SURVEYS

Newspapers, magazines, television stations, and manufacturers often conduct polls to determine people's opinions. On the basis of contacting no more than a few thousand people, a candidate is considered unbeatable, a television show is canceled, or a new soft drink is introduced to the consumer market. Obviously, such surveys can have a tremendous impact on society.

Surveys can also be the basis for your science fair project. You can survey almost anything, from the types of bacteria growing in your school to the kinds of dogs living in your town. Naturally, the larger in scope the survey, the more thorough you must be in your planning. Be sure the size and composition of the surveyed population are adequate.

You can make valid conclusions from relatively small sample sizes only if the individuals being surveyed are a representative cross section of the population. If only a few individuals are surveyed, any conclusion you reach regarding the total population might not be valid.

Just think about what would happen if you asked ten friends what sports they liked or songs they enjoyed. Your friends' responses would probably agree with your own. Chances are, however, that many other people in your community would not agree with you. If you survey only your friends, your results will not reflect the opinions of the community as a whole.

A survey may lead to other scientific processes that can become part of your project. Before drawing any conclusions from your survey, you may need to use mathematics to analyze your results. If you do, learn some *statistics*—a branch of mathematics that deals with the collection, analysis, and interpretation of data. Check the section "Analyzing Your Results" in Chapter 7 for a brief discussion of a few statistical procedures. Your math or science teacher can help you determine which statistical tests will be most useful for analyzing the results of your survey.

Graphs, data tables, and written reports can also be part of your survey. The results, if conducted over a period of time, may even enable you to make a scientific prediction about future trends. Predictions concerning human behavior are often made and can yield interesting results.

You may want to survey the patterns of eating behavior among students in different grade levels. Do types of food, number of meals and snacks eaten, preferences for fast foods, and diet regimens vary among different ages or genders? If you survey patterns of eating behavior, can you make any prediction as to what type of fast food will be the most popular next year?

In addition to eating habits, many other behavior patterns are constantly changing among students. Working with a teacher, you could develop a questionnaire that explores patterns of social behavior in your school. Do all students enjoy the same sports, watch certain television shows, or have similar hobbies? Remember that many science fairs have regulations regarding the use of human subjects in projects, so be sure to check these rules.

REPEATS FROM THE PAST

Some projects repeat a classic experiment or demonstration found in a textbook, lab manual, or book such as *Great Science Experiments: Twenty Experiments that Changed Our View of the World* or *Experiments in Chemistry*. See the bibliography for the names of the authors and publishers of these books.

Like models and displays, such a project may involve little or no creativity. Try to add something original, perhaps demonstrating some practical application. Take the classic experiment performed by the chemist Jacques Charles in 1787. He found that all gases expand by the same fraction of their original volume when they are heated over the same temperature range.

If you check a chemistry textbook, you'll discover that you could repeat his experiment with a piece of glass tubing and a drop of mercury. But to be more creative (and avoid working with mercury, which is toxic), you might demonstrate the application of Charles's law to hot-air balloons. Can you design a balloon that operates more efficiently when the heated air expands according to Charles's law?

Perhaps you can approach a classic experiment from a different angle. Review the work of Gregor Mendel. His experiments with pea plants in the 1860s were the first carefully planned investigations in genetics. All you need to repeat his experiments are pea plants, a place to grow them, and lots of time. Mendel took years to perform his genetic studies. He analyzed seven hereditary characteristics in pea plants, but you can save time by studying just one. You could save even more time by writing a computer program that summarizes Mendel's work.

By the way, notice how simple equipment and a few materials (glass tube, drop of mercury, and pea plants) were used to make major scientific discoveries. Your project may require fewer materials than you think. Ariel Ronneburger required only a few plants and some bacteria for a project that examined a controversial issue that has been debated for many years by scientists.

For her award-winning project, Ariel Ronneburger decided to investigate a controversy that has been going on for years in science. So don't hesitate to look to the past for ideas.

Ariel decided to test whether vitamin C could deter the growth and spread of cancerous tumors. Her research of the literature, including a search of the Web, revealed that scientists have conducted numerous experiments. Some of these experiments seemed to support the idea that vitamin C can benefit a variety of conditions—from the common cold to cancer, while others did not.

Ariel then divided plant seedlings into two groups. Both groups were injected with bacteria that cause tumors to form in plants. Both groups received identical treatments except for one factor—the experimental group received 500 milligrams of vitamin C daily, while the *control group* did not.

After 3 weeks, Ariel noticed that all the plants had developed tumors. However, those that had been given vitamin C had significantly smaller tumors. Ariel also observed that these plants did not develop secondary tumors at other sites along their stems. Ariel's results can now be added to those that have made up the long history examining the benefits of vitamin C.

ORIGINAL INVESTIGATIONS

Conducting a classic experiment might pave the way for the final type of project: an original investigation. If you repeat Mendel's work with garden peas, you may become so intrigued by heredity that you'll decide to do a project in genetics on a plant species about which little is known. Similarly, Charles's work can be the basis for a space-shuttle project, such as: How does the absence of gravity affect the behavior of gases when they are heated?

Keep in mind that an original investigation is often the most difficult kind of project to plan and conduct. An investigative project usually requires more library research, critical thinking, and laboratory work. If you are interested only in perfecting your skills in model building or assembling a display, then a project from one of the first four categories would be more appropriate.

However, if your ambition is to understand how scientists often approach problems, then you should conduct an original investigation. In this case, you will have to think of an original idea, perform experiments, collect information, and arrive at a conclusion. Because this type of project requires more work, most entries at science fairs fall within the first four categories.

Don't hesitate to carry out an original investigation simply because you feel the project may be overwhelming. In fact, an original investigation can be less involved than a noninvestigative project. Consider what Anokye Blisset did in her investi-

As Anokye Blisset proved, you don't need a lot of equipment or supplies to carry out an original investigation.

gation, titled "Will String Beans Germinate Better in Sand or Soil?"

Anokye divided bean seeds into two groups. One group was planted in sand, the other in potting soil. After treating both groups the same way for a month, Anokye found that the seeds in both groups eventually grew to the same height and appeared equally healthy. The seeds grown in sand took just a little longer to attain the same appearance as those germinated in soil.

However, you may decide to carry out an investigative project that involves more materials, requires more planning, and poses more questions. Consider the project carried out by Mohit Dugar. The title of Mohit's project is "Intron Sequencing of the Glia Maturation Factor Gene Using Long Polymerase Chain Reaction." Mohit spent 6 weeks of the summer following his junior year in high school at a university research laboratory

During that time, he became actively involved in a project that the staff had undertaken earlier. Their previous efforts had isolated a gene that exists in the nerve cells of various animals, including humans. The gene is responsible for producing a protein that has been shown to be an effective anticancer drug. Knowing the exact sequence of the components that make up this gene would be a valuable contribution to understanding the genetics of cancer.

Mohit spent 2 weeks learning the procedures he would need to carry out his sequencing project. Part of the gene had been sequenced by the laboratory staff, and seven gene fragments remained to be sequenced. Mohit discovered that it took approximately 40 hours to sequence just one fragment because of the time-consuming procedures he had to follow.

In addition, he had to use several pieces of equipment that are commonly found in research laboratories—but not in high school labs—including polymerase chain reaction machines and gel electrophoresis. In addition, he had to work with the computer people at the university to complete the final step—getting the actual sequence for each fragment, based on the lab results he obtained from five separate trials. Before he left the lab, Mohit was able to determine the sequence of four of the seven fragments. Since his return to school, the laboratory staff has finished sequencing the remaining three fragments.

For his investigative project, Mohit had to spend his summer vacation away from home, accept the responsibility for being part of a research team, learn how to use sophisticated equipment and carry out time-consuming tasks, and spend 40 hours sequencing a single, relatively small fragment. If you feel

Some original investigations, such as the one Mohit Dugar undertook, may require equipment and supplies that you can find only at a research laboratory or university. Mohit (left) is talking to a science fair judge about his project.

uncertain about making this kind of commitment, choose a less challenging investigative project. You may even want to reconsider doing a noninvestigative project.

SOME OTHER REASONS FOR DOING A NONINVESTIGATIVE PROJECT

Before tackling an investigative project, consider some valid reasons why one of the first four categories may be more suited to your needs. If you are planning to enter your first science fair, a display or model may be a better way to become acquainted with the steps necessary to complete a project. Constructing a model can be a learning experience and a way of "getting your feet wet" in all aspects of a science project. Also, if your background in science is not especially strong or extensive, you may be more comfortable with a survey or a display than with an original investigation.

A noninvestigative project can also create an awareness and appreciation for science, revealing its nature as fully as an investigative project. You can appreciate the importance of a noninvestigative project by examining the work of Charles Darwin, especially his studies done on the Galapagos Islands.

Evolution might very well be the first word that comes to mind when you think of Darwin. His theory of natural selection is one of the major unifying themes in biology, applying to all types of organisms. Yet Darwin also explored heredity, the growth of plants, the formation of atolls, and many other scientific areas. However, many of Darwin's important discoveries and valuable contributions did not come from experiments, but rather from his ability to observe nature and ask questions.

Darwin spent 5 years as a naturalist aboard a ship, traveling around the world. His curiosity led him to explore the mysteries hidden from the casual observer or the uninterested spectator. He collected large numbers of specimens, from barnacles to birds, and then proceeded to describe them in detail, marveling at their great diversity. Darwin's curiosity led him to develop an interest in science that grew in both scope and depth. Much of his work did not depend on original investigations, yet all scientists recognize the value of his studies.

Like Darwin, you may want to explore the world around you. You could construct a model of the moon, demonstrate how tides form, or collect and classify organisms that live in your local pond. While these projects do not involve the experimental procedures usually followed in an investigative project, they do require you to observe, explore, question, and describe just as Darwin did. This type of project might even lead you to ask some question that can be answered only with an original investigation.

PROJECT CATEGORIES

Examine the following titles to get an idea of the wide range of possibilities for projects. Included are ideas for models, displays, surveys, and both repetitive and original investigations. These titles are from projects submitted at a recent International Science and Engineering Fair, the largest science fair in terms of states and countries represented. The ISEF includes projects in behavioral and social sciences, computer science, engineering, earth and space sciences, mathematics, medicine and health, and zoology.

Moods and Music
Optical Illusions: Do You See What I See?
Morality in Children
Chemistry and the Personal Computer: Does the Computer Enhance Learning?
A Smart Pill in Our Future?
Artificial Embryo
Computers and Society
Teaching Sign Language with a Computer
Motion of the Ocean
Can an Apple Computer Recognize Speech?

Magical Numbers and Series
Soap-Film Math
Physics of Twisting and Somersaulting
What Is the Best Load for a 7-mm Magnum Rifle?
The Real Reason Airplanes Fly
Extending the Life of Ordinary Light Bulbs
Artificial Intelligence: Logic of the Mind

ANOTHER LIST

If you decide to do an original investigation, there are numerous areas you can explore. Examine the following list for ideas. You can change any title to match your interests.

Three-Year Study: Effect of Color on Mice
Effects of Stress on Alcohol Consumption in Mice
Effects of Colors and Word Ordering on Memory
Biochemical Changes in Seeds after Germination
Optimizing Hydrogen Photoproduced by Marine Algae
Meat Tenderizers: Study of Enzyme Activity
Comparison of Sugar Content in Soft Drinks
Comparative Study of the Fertilizer of Some Marine Algae
Factors Affecting the Growth of Geranium Cells
Effects of Microwaves on Seed Germination
Using Electricity to Control the Direction of Root Growth
STOP! You May Be Using the Wrong Detergent!
Comparative Studies of the Deterioration of Vitamin C in Various Samples
Design, Construction, and Efficiency Analysis of a Solar Pond
Effects of Outboard Motor Exhaust on Selected Marine Organisms
Five-Year Acid Rain Analysis through Various Studies
Can Dandelions Be Used to Measure Lead Pollution in Soil?

Effects of Thirst Quenchers and Exercise on Blood Pressure
Vitamin D and Its Effects on Rats
Effect of Bee Toxin on Rheumatoid Arthritis
How Antiseptics Affect Bacteria
Salad Bowl Microorganisms
Sage: A Natural Preservative
Food Preferences of Wild Birds
Effects of Magnetism on the Regeneration of Planaria
How Alcohol Affects Chicken Embryos
Pollution: Effects of Copper on Oysters
Effects of Ultrasound on Growth and Development
Tumor Inhibition by Hydrocortisone
Relative Effectiveness of Stain-Removing Toothpaste
Are Human and Animal Blood Compatible?
Improving the Aerodynamic Efficiency of Human-Powered Vehicles
Computerized Transportation System for the Handicapped
Synthesis of Anticancer Drugs from Platinum Complexes

As you can see from this list—anything goes. Your project can be original or repetitive, investigative or nonexperimental, practical or theoretical, mathematical or musical, pictorial or written. No matter what format you choose, your project should be a pursuit of something that has aroused your curiosity and interest. As you work on your project, you will discover the spirit of science.

CHAPTER 3

Planning for Your Project

Before beginning work on your project, pay attention to some details that can save you work and time. These involve developing plans to refine your topic, locating sources of equipment and materials, and working out a time schedule for each phase of your project.

NARROW YOUR TOPIC

If you look again at some of the project titles listed on pages 25 and 26, you will recall that some were too vague and gave you no idea where to begin. Once you have selected your topic and type of project, refine it so that you know exactly what you will be doing. This refining process is especially important if you plan to carry out an original investigation.

The library can be your best source for refining your proposal. Your detective skills will be important in tracking down and digging out the necessary information. Don't hesitate to ask the librarian for help. Even Holmes required Dr. Watson's assistance in solving some of his cases.

Let's say you want to do a project with the title "What Are the Effects of Acid Rain on Town Pond?" You'll discover that answering this question probably requires a team of experts, several years of work, and a considerable amount of money and equipment. Town Pond is inhabited by a wide variety of plants and animals of various ages. Each organism has unique biological activities, behavioral patterns, and seasonal adaptations.

Use the library to get more specific information about acid rain. You might come up with a plan to investigate the effects of acid rain on the photosynthetic rate of a specific plant or on the reproductive capacity of a particular animal found in Town Pond. A plan with either question is manageable and can yield some results within a reasonable amount of time.

You do not have to work out all the details now, but you should refine your topic until you can formulate a specific statement about what you plan to do. Setting your proposal down on paper will help you focus on your plan and also allow others to offer suggestions and advice. You are not committed or limited to your proposal; you can change it as you work through the final plan for your project. As with any detective or scientific investigation, you can modify your proposal to accommodate unforeseen circumstances or problems.

RECORD WHAT YOU READ

While searching through the library or on the Web, have index cards available. Organize important information on these cards. For a book, each card should contain the author's name, the title, the publisher, and the publication date. For a magazine or journal article, list the author's name, the title of the article, the name of the magazine or journal, the date of publication, and the page numbers of the article. For an encyclopedia reference, include the name of the encyclopedia, the publisher, the volume number, the date of publication, and the page numbers.

Each card should also have a short summary of the main points covered in the book or article. Write down this information for every reference, even if you don't see how you might use it for your project. You may discover later that a reference you considered useless turns out to be quite helpful.

Arrange all your index cards in some logical or meaningful pattern. For example, if you are researching antibiotics and bacteria, place all the cards with information on growing bacteria cultures into one group, those describing the various types of bacteria into another, any on the mechanism of antibiotic action into a third, and so on. At this point, set aside (but do not discard) any references that seem useless. Later, you may find that you need a particular reference, so save all your cards until your project is complete.

PLAN YOUR EQUIPMENT NEEDS

Next, make a list of the equipment, supplies, and materials you will need to conduct your project. Your list should be as specific as possible.

Compare the equipment in the following two lists:

List 1
assorted glassware
incubator
growth media
bacterial cultures
antibiotics

List 2
100 150-x 15-mm glass petri dishes
100 20-x 150-mm Pyrex test tubes
1 culture incubator with 30°C to 65°C range
50 g of nutrient agar
2 cultures each of *E. coli*, *B. subtilis*, and *S. lutea*
1 kit of antibiotic disks containing penicillin, erythromycin, neomycin, and streptomycin

List 1 is too vague; you have no idea what size the test tubes must be or how much agar will be needed to grow the bacteria. List 2 is more detailed and provides a clearer idea of the purpose of your project. You could easily locate all the items in a scientific supply company catalog. Knowing the type and quantity of each item, you can calculate the cost of purchasing it from a scientific supply company. Table 2 on pages 58 and 59 lists companies you could use to order equipment and materials.

Your science teacher probably has the catalogs of some of the companies in Table 2. Look through them to get ideas about what equipment and supplies are available. To avoid delay when you are ready to begin work, order the items as early as possible.

Because most of these science supply companies do not sell directly to individuals, you will have to place your order through your school. Speak with your science teacher. If you give your science teacher the money, he or she may be willing to place the order and give you the items when they arrive. If such arrangements are not possible, look closely through the Edmund Science Company catalog. They will sell many items directly to individuals.

PLAN YOUR BUDGET

Determine the total cost for your equipment and materials, but don't despair if the amount is too high. You can cut expenses in a number of ways. You may be able to find most of the materials you need right in your school. Check to see what is available in the science, mathematics, and computer departments.

If your school doesn't have something you need, check local hospitals, medical offices, junkyards, lumberyards, industrial sites, research facilities, or any other location where a donation may be obtained. For instance, the cost for a temperature-controlled incubation chamber may be too high, but you might be able to build one instead. You could use a cabinet from a junkyard, a thermostat from a physics laboratory, insulation from a lumberyard, and spare parts from a hospital.

TABLE 2 SCIENTIFIC SUPPLY COMPANIES

Company	*Address*	*General Information*
Carolina Biological Supply Co.	2700 York Rd. Burlington, NC 27215 1-800-334-5551	Don't let this company's name fool you. They carry equipment and materials for projects in biology, physics, and chemistry.
Central Scientific Co.	3300 CENCO Parkway Franklin Park, IL 60131 1-800-262-3626	They can provide supplies for most projects and carry reference books.
Connecticut Valley Biological Supply Co.	82 Valley Road Southampton, MA 01073 1-800-628-7748	They focus mainly on biology with a limited supply of chemical reagents.
Edmund Science Co.	101 E. Gloucester Pike Barrington, NJ 08007 1-609-573-6250	They carry a wide range of equipment including microscopes, weather instruments, motors, and pumps.
Fisher Scientific Co.	711 Forbes Avenue Pittsburgh, PA 15219 1-800-766-7000	They have a very comprehensive catalog.
Frey Scientific Co.	100 Paragon Parkway Mansfield, OH 44901 1-800-225-3739	They offer equipment and supplies for projects in the life, physical, and chemical sciences.

Company	*Address*	*General Information*
Flinn Scientific	P.O. Box 219 Batavia, IL 60510 1-800-452-1261	They currently specialize in chemistry, and are in the process of expanding their biology offerings.
Nasco	901 Janesville Ave. Fort Atkinson, WI 53538 1-414-563-2446	
PGC Scientifics	P.O. Box 7277 Gaithersburg, MD 20898 1-800-424-3300	
Sargent-Welch Scientific Co.	P.O. Box 5229 Buffalo Grove, IL 60089 1-847-459-6625	
Science Kit and Boreal Labs	777 E. Park Drive Tonawanda, NY 14150 1-716-874-6020	
WARD'S Natural Science Establishment	P.O. Box 92912 Rochester, NY 14692-9012 1-800-962-2660	

Don't hesitate to ask people if they might have something to donate, so long as you explain the reason for your request. In most cases, they would prefer to see an item used rather than sitting in a storage closet waiting to be tossed. Be sure to write a thank-you note and acknowledge any contribution to your project.

If your school offers classes in woodworking, small-engine repair, basic electronics, or similar subjects, you may be able to build equipment that you can't afford to buy. If you tell an industrial arts teacher what you need, he or she may be able to help. The teacher may even be able to suggest ways to improve your design.

You can also get plans, ideas, and suggestions for building equipment from books such as *Projects for the Amateur Scientist* and *Entertaining Science Experiments with Everyday Objects* and from magazines specifically geared to science teaching such as *The Science Teacher*, *Science and Children*, and *The American Biology Teacher*. Publications aimed at the home hobbyist include *Popular Mechanics* and *Home Mechanix*.

You might be able to substitute something for an item you can't buy or locate, or just to save some money. Look around at what is available in school or at home and see if anything can be adapted to fit your needs. An assortment of jars can substitute for a supply of beakers, measuring cups for graduated cylinders, and an aquarium pump for an aerating system.

Most science fair judges give credit for ingenuity and admire projects that show creativity in overcoming an obstacle, such as a missing piece of equipment. A project showing some creativity may get more recognition than one in which the equipment was purchased. Judge for yourself by looking at how the following two projects reflected resourcefulness and creativity.

SOLUTIONS TO THE EQUIPMENT PROBLEM

For her project, Sara Henck wanted to test how various brands of cigarettes affected heartrate. For her experimental subject, she chose daphnia, a small invertebrate that lives in fresh

Sara Henck found many of the items needed for her project right in her own house. She used a bicycle pump to deliver cigarette smoke to her experimental subjects.

water. The problem she faced was how to expose the daphnia to controlled amounts of cigarette smoke. One simple solution was to purchase some expensive pieces of equipment.

Sara decided against this plan. Instead, she decided to look around her house for ideas. She found some pieces of plexiglass and came across a bicycle pump. Using the plexiglass pieces, Sara built a "smokehouse" in which she could trap the smoke from the various brands of cigarettes she was testing. To prevent the smoke from escaping, she simply placed her smokehouse in a pool of water contained in a baking tray. She rigged up a bicycle pump to deliver controlled amounts of cigarette smoke to the daphnia.

Sara's ingenuity paid off; she received an honors award for her project at a regional science fair. By the way, her results showed that one brand of cigarettes had a more significant effect than the other three in lowering the daphnia's heart rate.

Another project that involved some common household items was done by Jeremy Hwang. His project was titled "The

Jeremy Hwang used empty plastic soda bottles as environmental chambers for growing the plants he used in his project.

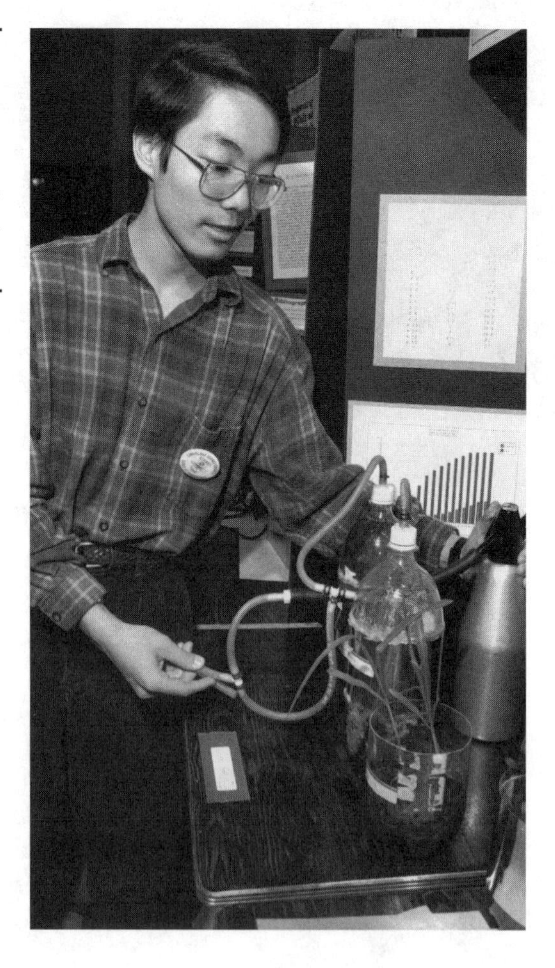

Effect of Carbon Dioxide on the Growth Rate of Plants." Jeremy wanted to test how carbon dioxide affects the growth of plants that carry out photosynthesis in different ways. The problem was to maintain the plants in a controlled environment where Jeremy could introduce a known quantity of carbon dioxide gas. Environmental chambers are available, but they are expensive.

Searching for an alternative, Jeremy decided to build a simple chamber out of a 2-liter plastic soda bottle. He placed each type of plant in a different container and used a carbon dioxide cartridge from a seltzer dispenser to supply the gas. The carbon dioxide gas passed from the dispenser, through rubber tubing, into the plastic chambers containing the plants. Jeremy found that both plants grew equally well. But more importantly, he found that he could complete his award-winning project with some simple things he had around his home.

PLAN FOR SAFETY

As you plan your project, be aware of any equipment that could be hazardous. Equipment that could injure someone or pose a safety threat is prohibited from all science fair competitions. In most cases, displays may not include disease-causing organisms, dangerous chemicals, combustible materials, explosive gases, open or concealed flames, radioactive sources, or unsafe equipment. You can use these items only if they are necessary to carry out your project.

If you get permission to use such items, check with your teacher or a professional for information about safety precautions. Take the necessary safeguards to prevent any accidents. Many projects require some degree of caution. Examples include any project that requires the use of chemical reagents, open flames, or combustible materials. The only way to avoid these materials is to plan a project limited to a display of pictures and photographs or a written report based on library readings.

PLAN YOUR TIME

Finally, plan a time schedule. A model may require only a few hours of work to complete, while an original investigation may take hundreds of hours. Be aware of the time required before you start your project. Talk to your teacher or someone who has done a project similar to the one you're planning to get an idea of how much time is involved.

Divide your project into small segments and set a due date for each one. In this way, you'll get a feeling of accomplishment as you meet the deadline for each step. For example, if you plan to build an original model, plan the dates for selecting your topic, refining your idea, gathering the equipment and materials, finishing the model, and preparing any written reports. See Table 3. If you plan to enter your model in a science fair, have everything ready by that date.

TABLE 3 PLANING A SCHEDULE FOR YOUR PROJECT

Step	*Time (in days)*	*Due Date*
Selecting topic	7	October 15
Refining topic	15	November 15
Gathering equipment and materials	5	December 1
Conducting experiments	40	April 1
Analyzing results	7	April 15
Preparing report	10	May 1
Preparing presentation	1	May 15

Once you have refined your topic, located sources to obtain equipment and materials, and worked out your time schedule, your plans are completed. Now get ready to begin the actual work.

CHAPTER 4

Scientific Processes

If you plan to do an original investigation for your project, you should be familiar with the processes scientists use in their work. You have read about some of these processes in earlier chapters—making observations, collecting data, and researching the literature. Table 4 on the next page includes a number of processes used by scientists. Exactly which processes they use depends on the type of work they are conducting.

For example, if scientists are conducting a survey, they collect, organize, interpret, and graph data in order to draw conclusions. If they are testing the effect of a new drug on a particular disease, they form a hypothesis, conduct an experiment, analyze their results, and then draw conclusions. If you use these processes to do your investigations, you won't be guaranteed success, but you will be less likely to fail or give up hope of ever completing your project.

TABLE 4 SCIENTIFIC PROCESSES

Making observations	Making models
Forming hypotheses	Reviewing the literature
Conducting experiments	Making inferences
Recording data	Drawing conclusions
Making measurements	Verifying results
Modifying hypotheses	Communicating findings
Conducting interviews	Sharing information

FORMING YOUR HYPOTHESIS

When you begin planning your science fair project, the first thing you must do is come up with a *hypothesis.* A hypothesis is nothing more than an educated guess, proposing a possible solution to the problem or question raised by your original investigation. Keep several factors in mind when forming your hypothesis.

You should base your hypothesis on the information gathered from your research, particularly any library readings and on-line searches you may have done. Your hypothesis should take into account the results of any experiments or observations mentioned in these sources.

Your hypothesis should be clear and brief, providing a good idea of what you plan to do without being too wordy. State your hypothesis in one sentence: "Small amounts of caffeine will speed up the growth rate of earthworms." From reading just this one sentence, anyone can tell what you intend to explore in your investigation.

Remember that you must be able to test your hypothesis. If you are planning to investigate whether caffeine speeds up the growth rate of earthworms, you can place the earthworms in soil samples containing different amounts of caffeine and compare their rates of growth. You cannot, however, form a hypothesis suggesting that caffeine will make the worms feel better, because you cannot perform any scientific tests to measure their feelings.

Once you have formed your hypothesis, prepare a brief statement explaining your purpose: "Caffeine is a drug that acts as a stimulant, speeding up many metabolic processes. The purpose of this project is to determine if caffeine will cause earthworms to grow at a faster rate."

Remember, your hypothesis is only an assumption or an educated guess. If you have done extensive research to refine your topic, your hypothesis may be based on firmer ground, especially if you have a good deal of background information supporting your position. Nonetheless, your hypothesis is still only an educated guess. You may find that the results of your investigation do not support your hypothesis. For example, you may find that caffeine has no effect on an earthworm's growth rate. You may even discover that caffeine inhibits the growth rate of earthworms.

Do not consider your project a failure if your investigation does not confirm your hypothesis. If this happens, *only* your hypothesis is discarded. The results of your investigation are still valid and can be used as part of your project. In fact, these results might be helpful in providing you with ideas and clues for future projects.

CONDUCTING YOUR EXPERIMENT

The next step is to design your experimental procedure. Outline the steps you will follow in attempting to reach some answer or conclusion about your hypothesis. Make the experimental design as simple as possible. The best scientific investigations are often simple in nature and direct in approach. The more complicated the design, the greater the chance for error. In addition, you may never finish your project if it is too complex.

There are two types of experiments: qualitative and quantitative. A *qualitative experiment* can be conducted through careful observations without getting involved in measurements or statistical analysis. In a qualitative project, there may be no need to collect data by recording time, for example, or measuring changes in volumes.

A *quantitative experiment* involves measurements and collecting numerical data. If you plan to enter your original investigation in a science fair, be aware that the judges often look to see if you have included quantitative information wherever possible. Carefully examine your experimental design to see where you can make some measurements and record numerical data. Although you may think your project does not require any measurements, a longer look may reveal some quantitative aspects.

For example, a project examining the food preferences of a particular invertebrate seems qualitative: The animal prefers lettuce to liver. Nevertheless, you should compile a data sheet, indicating the number of times the organisms responded to each type of food, how long each response took, and the length of time spent consuming the food.

Similarly, a project exploring the adaptations of insects to pesticides may not appear to be a quantitative investigation. You may simply observe whether the organisms live or die. However, you can measure and record the amount of pesticide used, the survival rate at various concentrations, and any abnormal growth patterns. You could then make a more specific or quantitative conclusion about the effects of the pesticide. Perhaps a particular concentration of pesticide is needed to produce an abnormality in growth.

In any case, look closely for ways in which you can incorporate quantitative data. As an example, consider how Jordan Adler, Jonathan Ells, and Jeffrey Hardgrove incorporated quantitative data into what appears to be a strictly qualitative project based on its title—"Recycling Shower Water for Use in Toilets."

The students designed a system whereby all the water used for showers is collected in a holding tank stored in a basement. Normally such water passes directly out through the sewage system. The holding tank has an overflow valve to release excess water and another valve to let in water from the main supply line if the level in the tank drops too low. Whenever a toilet is flushed, an on-demand pump on the holding tank is activated and sends water to the tank on the toilet.

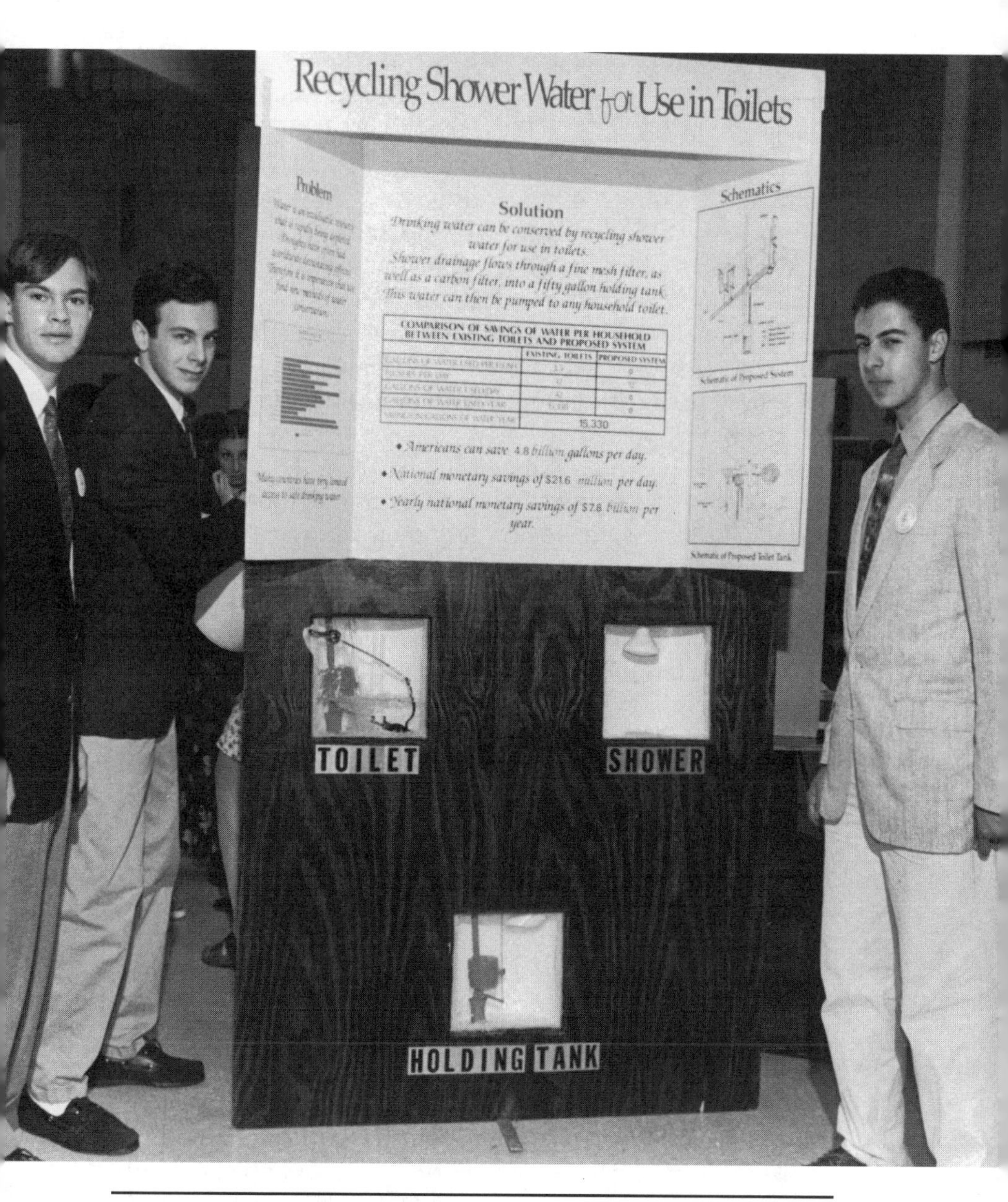

Jordan Adler, Jonathan Ells, and Jeffrey Hardgrove saw how quantitative data can really impress the judges.

But in addition to designing the recycling system, Jordan, Jonathan, and Jeffrey also calculated how much water and money could be saved. They determined that an average household flushes a toilet *twelve* times per day. Each flush requires 3.5 gallons of water, thus using 42 gallons per day. Over a year, their recycling system would save a household 15,330 gallons of water. They calculated that this translates to 18 billion liters 4.8 billion gallons of water for all the households in the United States. Based on cost figures they obtained from several sources, they estimated that their recycling system represents a nationwide savings of $20 million per day and $7.3 billion per year.

For their project, Jordan, Jonathan, and Jeffrey won the energy award at a regional science fair competition. Obviously, the judges recognized not only their innovative recycling model but were also impressed by their data, especially their projected savings estimate. Notice that their calculations were based on quantitative data expressed in gallons, a unit that makes sense because household water consumption is measured in gallons. However, any data you collect in conducting your project will likely be recorded in metric units.

THINK METRIC

Whenever possible, all measurements in your quantitative investigation should be in metric units. Scientists throughout the world use the metric system to record length, volume, mass, and temperature. The English units of measurements are equally precise, but they are more cumbersome to use when trying to convert from one unit to another.

The metric system is preferable because the units are based on multiples of 10, so converting from one unit to another is easy. Just move the decimal point a certain number of places to either the left or the right. For example, 6.2 millimeters equals 0.62 centimeter, and 0.024 kilogram equals 24 grams. In metric there is no need to divide by 12, as when converting inches to feet, or to multiply by 16, as when changing pounds to ounces.

All your equipment and lab apparatus should be calibrated in metric. Do not use a cylinder or a thermometer graduated in English units with the idea of making all the conversions at the end of your investigation. The length of time required to perform all the conversions might be discouraging. Use metric measurements from the start so that your frame of reference corresponds to the system used by scientists throughout the world. Even if you have had little or no experience with metric units, you'll find they are easy to use.

ANALYZING YOUR RESULTS

The third process is analyzing the data you have collected in your experiments. Appreciate the importance of data by heeding Holmes's warning as given to Dr. Watson: "I have no data yet. It is a capital mistake to theorize before one has data.

If you keep all of the data and notes related to your project in a single notebook, you won't have to worry about losing information.

Insensibly, one begins to twist facts to suit theories, instead of theories to suit facts." Like the great detective, you must collect sufficient data to explain the facts uncovered in your project.

Do not record the data on pieces of scrap paper or in an assortment of notebooks used for other purposes. Keep a notebook exclusively for your project data. You may want to take photographs or make drawings and diagrams of various stages of your experiments, especially if your project involves qualitative descriptions such as changes in color or patterns of behavior. In any case, do not attempt to commit results to memory; you'd be surprised how quickly any mental records of data fade. Placing confidence in your ability to memorize will only cause you to lose data, some of which may be important pieces of the final solution.

DRAWING YOUR CONCLUSION

With your data in hand, you have reached the last step: drawing a conclusion. Your data will either support or reject your hypothesis. Without a conclusion, your project will have failed to fulfill its purpose, because the whole reason for your investigation was to answer a question or solve a problem. Don't spend so much time carrying out your investigation that you rush through your data and fail to arrive at a conclusion. Science fair judges will be looking for a conclusion to any investigative project. Without a concluding statement, your project will be considered incomplete.

Your conclusion must come directly and solely from the data in your notebook. If you cannot arrive at any conclusion with your data, you may not have finished your experiments. You may have to perform more tests or check out some different approaches. Once you obtain a sufficient amount of data, make your conclusion. Be clear and concise. Don't hesitate to present *all* the conclusions your data can support. Do as much with the data as possible. You may want to construct charts or tables that clearly show your conclusions and interpretations.

On the other hand, don't reach any conclusion that is not justified by your data. If you claim something that is unsupported by the results of your investigation, others will discover this discrepancy when they examine your data. You can point out some possible findings of your investigation, but emphasize that your results are only suggestive and not conclusive in these areas.

Again, don't worry if your data fail to support your hypothesis. You cannot change the data, but you can modify or even change your hypothesis and perhaps suggest a new set of experiments to test it. An insight into future possibilities would add to your credentials as a competent and thorough investigator.

Consider how Carolyn Pastor had to modify her hypothesis when she realized that her data did not support her original hypothesis. Carolyn's project, titled "Efficient Fan with Increased Air Velocity," was based on the hypothesis that increasing the surface area of the fan blades would both increase the velocity of air moving across the blades and decrease the fan's consumption of energy.

To test her hypothesis, Carolyn prepared twenty sets of fan blades, each with a different amount of surface area. To vary the surface area, Carolyn made dimples on the fan blades. Half of the blades had dimples made by drilling tiny holes into the surface. The other half had dimples made by using a glue gun to create tiny elevations on the surface. The number of dimples also varied among the twenty fan blades that Carolyn prepared.

She measured the velocity of the air that moved across the blades with the help of a draft meter. Carolyn used an ammeter to measure the number of amps the fan was using, so she was able to determine the cost of operating the fan. Carolyn discovered that increasing the surface area of the fan blades was beneficial, both in terms of increasing air velocity and decreasing costs, up to a certain point. Increasing the number of dimples beyond this point actually made the fan move the air less efficiently, both in terms of velocity and cost.

After analyzing her data, Carolyn Pastor had to modify her original hypothesis. In some cases, the data may result in a hypothesis being totally discarded. Yet the project can still be a winner because judges will recognize and reward originality, effort, and use of scientific processes.

Thus, Carolyn had to modify her original hypothesis to state that increasing the surface area of fan blades only within a certain range would make for a more efficient fan. The judges recognized Carolyn's ability to use a number of scientific processes and her thoroughness and creativity by awarding her highest honors at a regional science fair.

CHAPTER 5

Conducting an Experiment

If you decide to conduct an investigation for your project, be aware of several factors scientists consider in designing and carrying out experiments. Design your investigations so that you know what factor in the experiment is responsible for producing the results. In addition, collect accurate and sufficient data before making any conclusions. Finally, take precautions to reduce or eliminate sources of error wherever possible.

INCLUDE A CONTROL

An experiment often involves a *control*. A control sets up conditions in an experiment so that the investigator knows what factor is responsible for the results. Using a control is the only way to make sure that no more than one part of the problem or question is tested at a time.

For example, imagine that you are planning to investigate the effect of light intensity on the rate of photosynthesis in a freshwater plant. Your experimental design might involve submerging the plant in a test tube of fresh water and exposing it to light. When the light is turned on, the plant can photosynthesize.

Your hypothesis states that the greater the light intensity, the higher the rate of photosynthesis. From your biology course, you remember that a plant uses carbon dioxide and produces oxygen during photosynthesis. Your plan involves measuring the amount of oxygen produced by the plant as an indication of how much photosynthesis is occurring. The more oxygen produced, the higher the photosynthetic rate.

Your procedure involves varying the distance between the light source and the plant, and then measuring the amount of oxygen produced. Using several plants in different experiments over a period of 3 weeks, you discover that moving the light closer increases the photosynthetic rate—up to a certain point. But can you be sure that the greater light intensity was responsible for the higher photosynthetic rate? Perhaps the rate increase was caused by some other factor such as the availability of more carbon dioxide for photosynthesis. The higher rate of photosynthesis may have also been due to warmer temperatures on the day you conducted the experiment or to the amount of water in the test tube.

In any case, you could not conclude that greater light intensity was the only factor responsible for the increased photosynthetic rate. If you included a control for this experiment, however, you could make a valid conclusion about the relationship between light intensity and the photosynthetic rate.

To establish a control, you need two experimental setups identical in all respects except one. In other words, you need to compare what happens to a plant exposed to light and a plant placed in the dark. *All* other experimental conditions must be the same: the amount of water in the test tube, the size and shape of the plant, the amount of carbon dioxide, and the temperature.

The extent to which the light affects photosynthesis would be determined by subtracting the amount of oxygen produced by the plant placed in the dark from the amount produced by the plant exposed to light. Because all conditions except one are the same, you could prove a cause-and-effect relationship between the light source and the rate of photosynthesis: The greater light intensity must be the *only* factor responsible for any increase in the photosynthetic rate. With the use of a control for each experimental setup, you eliminate all other possibilities.

SOME IMPORTANT DEFINITIONS

You may come across some words or terms frequently used in investigative projects. Let's use the experiment testing the effects of different light intensities on the photosynthetic rate as an example. Any plants exposed to light make up the *experimental group*. The experimental group is the one exposed to the factor being tested, in this case light intensity. Those plants placed in the dark make up the control group. The control group is the one that is not exposed to the factor being tested in the experiment. In all other features, the experimental and control groups are identical.

In this experiment, light intensity is the *independent variable*. The independent variable is the factor that the experimenter is free to change at will. The photosynthetic rate depends upon the light intensity and is known as the *dependent variable*. The dependent variable is the factor that the experimenter is causing to change.

A control allows for only one independent variable to be tested at a time. Be careful to keep all other conditions the same. Plants must be randomly divided between the experimental and control groups—don't select the "healthier-looking" plants for the experimental group. Also do not favor one group with more water, cleaner test tubes, or additional carbon dioxide. If you do, more than one independent variable will be present, and any conclusion you make will be invalid.

In some experimental designs, the independent variable can consist of two or more factors that the investigator simultaneously varies. For example, if you design an experiment to establish the optimum soil condition for growing tomato plants, one procedure might involve varying the amounts of sand, clay, and humus in each pot. One mixture might contain equal amounts of each ingredient, another mostly humus, a third mostly sand, and so on. In this experiment, the independent variable is the chemical composition of the soil. All other factors, such as water, temperature, and light, must be the same.

THE NEED FOR CONTROL

To be sure you understand how to use a control, check the following list. For each investigation, explain how you would set up the experiment.

The Effects of X Rays on Seed Germination
Evaluating the Effectiveness of Various Food Preservatives
Which Mouthwash Best Prevents Bacterial Growth?

In every case, you will need two groups identical in all respects except one. To study how X rays effect seed germination, divide the seeds into two groups. The seeds in one group will be exposed to X rays, while those in the other group will not. To study food preservatives, treat one-half of a food sample with a preservative, and do not add any to the other half. To determine which mouthwash is most effective, add mouthwash to some bacterial cultures, but not to others. In each of these experiments, keep all other conditions between the two groups the same.

Meghana Bhatt recognized the importance of a control in her project entitled "The Role of Nitric Oxide on Cystic Fibrosis." Meghana conducted her project at a local hospital where several people with cystic fibrosis were being treated as outpatients. Her experimental group consisted of these

Meghana Bhatt's project involved both an experimental group and a control group. By comparing blood samples from the two groups, she concluded that people with cystic fibrosis have an elevated level of nitric oxide in the blood.

patients. Her control group was composed of people who did not have cystic fibrosis. Blood samples taken from both the experimental and control groups revealed that people with cystic fibrosis had elevated levels of nitric oxide in their blood. In addition, her project demonstrated that the cystic fibrosis patients have unusually high levels of a compound known as sulfidoleukotriene in their blood.

Analysis of her data also revealed a direct correlation between nitric oxide levels and sulfidoleukotriene levels—the higher the nitric oxide level, the higher the sulfidoleukotriene level. The reason for this correlation will be the basis for Meghana's next project.

Beth Topf did a project entitled "The Effect of Secondhand Smoke on Asthma and Allergies." Beth's project involved a survey of her fellow students. The experimental group consisted of those exposed to secondhand smoke, while the control group was comprised of those who were not exposed. Using a computer to analyze the responses to her questions revealed that those students exposed to secondhand smoke were afflicted with more allergies and more likely to have asthma than those who were not exposed.

Beth Topf included both an experimental group and a control group when she conducted her survey about the effects of secondhand smoke for her project.

KEEP ACCURATE RECORDS

As you conduct your experiment, you will collect data. Record all your data in a notebook used only for your project. The format of your notebook depends on the nature of your investigation. The figure on the next page shows how you might organize data obtained from a series of experiments that look at how light intensity affects photosynthesis.

Notice that the heading on the form contains spaces for the title of your project, the date, the number of times the independent variable was tested, and any additional information concerning the investigation. The columns are arranged so that the time elapsed and amount of oxygen produced can be easily recorded. Obviously, any format you design must include sufficient space for recording your test data. Set aside some additional space at the end of the form for general comments.

Title: Effect of Light Intensity on Photosynthesis Date: 3/15/86
Distance: 30 cm. Trial Number: 4
Comments:

Time (min.)	O_2 (ml.)	Comments

Prepare a data table in your notebook so that you can record pertinent information and observations.

Include these comments as part of your record keeping, no matter what format you use for recording data. You may observe something unexpected, get an idea to check something else, or think of a way of improving your experimental design. In any case, have a place to jot down this information before you forget.

Because you should carry out each experiment several times, make copies of your data form, using one for each experiment. No scientist can arrive at a valid conclusion from the results of one experiment. Before arriving at any conclusion, be sure to perform a sufficient number of tests.

For example, you may eventually test the effects of a light source placed at a certain distance ten times before reaching a valid conclusion. If you recorded 6 milliliters of oxygen produced the first time and 20 milliliters the second, you should not conclude that 13 milliliters—the average of the two results—represents the amount produced at that distance. However, if you performed ten trials and obtained measurements of 6, 20, 17, 9, 14, 10, 16, 12, 18 and 8 milliliters, then your average of 13 milliliters would be based on more substantial data.

OBJECTIVITY IS THE RULE

Be objective. Do not allow your personal feelings or opinions to interfere with your experiments. You may observe a result totally out of agreement with other data. Do not discard this result—there are no incorrect answers when conducting an experiment, only unexpected ones!

Although some results may not be understandable at the time, analyzing your records at a later date may reveal a possible explanation. Perhaps the lone result was caused by contaminated glassware, faulty equipment, or a math error. Even if your records do not provide any immediate explanation, an unexpected result may not be an error but a clue to some interesting discovery or accidental finding. Don't forget—penicillin was discovered when Fleming made an unexpected observation while working with bacterial cultures.

Any accidental finding points out that the scientific method is only a guideline for conducting an investigative project. Don't be afraid to branch out from your work or follow tangents suggested by your data. Obviously, you cannot be trained to make an accidental discovery, but you can remain alert to ideas, clues, suggestions, and leads provided by your data. Not following up on an interesting development may save some time. After all, the science fair might be rapidly approaching, and your project has to be completed on time. But what if that one unexpected result led to an interesting finding or an important discovery?

EXPERIMENTS ARE NOT FREE FROM ERROR

Errors can occur in any experiment. They can be caused by faulty equipment, human mistakes, or a host of other reasons. These types of errors do not detract from the quality of a good project. Don't hesitate to admit your mistakes; follow the example of Lara Ausubel. Lara explored the mechanism of action of verapamil, a drug used in the treatment of coronary artery disease.

Lara had read medical reports indicating that this drug probably prevented heart spasms by inhibiting the action of nerves that cause the smooth muscle of the heart to contract. She also learned that verapamil blocks the flow of calcium ions across the membranes of nerve cells and slows down the cilia of cells lining the trachea, or windpipe, of rats. For her project, Lara wanted to determine whether verapamil acted directly on heart cells.

Lara used single-celled organisms called protozoans as subjects. Because these organisms have cilia but no nerves, she could study whether the drug affected the cilia directly. She added serial dilutions of the drug to measured volumes of the protozoans. Lara set up appropriate controls and repeated each experiment several times. She observed the effects of the

Table 12. The Assay of cAMP in spring water and in *Paramecium* cultures, untreated versus treated with various concentrations of *Isoptin* ®, using the method of Tovey as modified by Chiang*

	cAMP levels in picomoles/ml.		
	Run 1	Run 2	Average
Spring water	0.00	0.00	0.00
Untreated Paramecia diluted with spring water (1:1)	0.17	0.23	0.20
Paramecia treated with *Isoptin* ® 1:750 (1:1)	0.17	0.28	0.22
Paramecia treated with *Isoptin* ® 1:700 (1:1)	0.09	0.27	0.18
Paramecia treated with *Isoptin* ® 1:600 (1:1)	0.20	0.17	0.19
Paramecia treated with *Isoptin* ® 1:500 (1:1)	0.10	0.21	0.16
Paramecia treated with *Isoptin* ® 1:500 (1:1)	—**	0.16	0.16
Paramecia treated with *Isoptin* ® 1:400 (1:1)	0.13	0.10	0.12
Paramecia treated with *Isoptin* ® 1:300 (1:1)	0.10	0.08	0.09
Paramecia treated with *Isoptin* ® 1:200 (1:1)	0.06	0.18	0.12
Paramecia treated with *Isoptin* ® 1:100 (1:1)	0.00	0.06	0.03
Paramecia treated with *Isoptin* ® 1:50 (1:1)	0.00	0.00	0.00
Paramecia treated with *Isoptin* ® 1:10 (1:1)	0.00	0.00	0.00
Paramecia treated with *Isoptin* ® 1:5 (1:1)	0.00	0.00	0.00
Paramecia treated with *Isoptin* ® 1:2 (1:1)	0.00	0.00	0.00
Paramecia treated with *Isoptin* ® (1:1)	0.00	0.00	0.00

* [^{3}H] Cyclic AMP Assay kit supplied by Diagnostic Products Corporation, Los Angeles, California

** Some of the ^{3}H cAMP was spilled during transfer, making analysis impossible

drug by making microscopic observations. The figure on the previous page is a reproduction of a chart Lara created to show the results of each step of her experimental procedure.

Lara observed that as the concentration of verapamil increased, all the cilia became paralyzed, suggesting the drug worked directly on heart cells rather than on the nerves that control it. She also discovered that this paralysis was caused by the drug blocking the conversion of a chemical known as ATP into another compound known as cAMP.

Again check Lara's chart for the culture treated with the 1:500 dilution. Notice that Lara could not report any results for run 1 because some of the cAMP was spilled during transfer—a human error. However, this in no way affected the outcome of her project.

Lara received highest honors at the regional science fair and was selected to enter the state competition, where she again won highest honors and received first prize for best biology project. The judges obviously recognized the scientific value of her project and totally ignored the minor mistake in one of her experiments.

HOW TO CONTROL SOME ERRORS

If a piece of equipment gives erratic readings, you can get it repaired or replaced. However, nothing can be done about the slight, built-in variations in balances, thermometers, and rulers, or about minor, human misjudgment when measuring. If you weigh the same object on ten different balances to the nearest 0.01 gram, you may not get the same measurement in all cases. If you record the temperature of a water bath with ten different thermometers, you may find slight differences.

Be careful not to allow errors to get out of hand. The accuracy of your data depends on how precise and careful you are when making measurements. The more accurate your data, the more solid your basis for making conclusions.

CHAPTER 6

Preparing Your Report

No matter what type of project you do, you will have to prepare a report summarizing your work. If your project centers around a model, pictorial display, or a repetition of someone else's work, your report will probably be less extensive than one describing an original investigation. Still, you can use your imagination to make your report interesting, perhaps even extraordinary.

For example, if you build a model, include drawings or plans showing the various stages of construction. A blueprint drawn to scale of the completed model can be a way of including some quantitative data in your report. You can explain technical problems encountered or scientific applications uncovered while assembling the model.

For a project centering on a pictorial display, your report can describe the scientific story behind the photographs, pictures, or drawings. If your project involves a repetition of a classic experiment, include suggestions for improvement or ideas for additional investigations.

If you have conducted a thorough and extensive lab investigation, you may face a problem shared by Dr. Watson, who wrote, "When I glance over my notes and records of the Sherlock Holmes cases . . . I am faced by so many which present strange and interesting features that it is no easy matter to know which to choose and which to leave."

Your data notebook may be filled like Dr. Watson's chronicles. Where do you begin and what will you choose to include in your report? No matter how you decide to proceed, be sure to type your report, using a word processor with a spell checker. Judges are likely to give low scores to a sloppy report, even if it represents an impressive effort.

START WITH A TITLE PAGE

Begin by preparing a cover or title page for your report. Don't underestimate the importance of a good title. Spend some time thinking about the title. A good one will grab everyone's attention. Don't be vague; if possible include both the independent and dependent variables tested in your investigation. Consider the following two titles:

A Burning Issue—The Effects of Retardants on Flammability
Computer Math

The first title is creative and gives some information about what the project covers. The second title tells you only that the project involves computers and math.

THE BODY OF YOUR REPORT

There are no absolute rules as to how you should write your report, but most papers are prepared according to the format used in scientific journals: brief summary of the work; introduction; explanation of the methods; results; conclusions, including the significance of the work; and bibliography.

A brief summary of a written scientific report is called an abstract. Although the abstract comes first, you may find it easier to write after you have finished your entire report. Because your abstract is probably the only part of your report the science fair judges will read carefully, be clear and concise.

Explain the purpose, procedure, and conclusion of your project in three or four paragraphs totaling 200 to 250 words. Choose your words carefully, because you want to arouse the reader's interest in your project. See how the following abstract accomplishes these objectives.

> The purpose of my project is to explore the virulence relationships among three species of the bacterial genus *Yersiniae*. The best-known species is an intracellular parasite responsible for bubonic plague. The other two are less virulent but can cause severe gastrointestinal disease in humans and animals.
>
> Research work indicates that the virulence of these species depends on genes located on the plasmids, pieces of hereditary material separate from the main bacterial chromosome. Little is known, however, about the role of plasmid genes in causing disease.
>
> These plasmids also produce a substance that agglutinates red blood cells. Since the production of this substance can be observed and quantitatively read, a study of this blood-clotting process could serve as a model for understanding how these bacteria cause disease.
>
> Significant variations of the blood-clotting substance were found among the three species, with the highest level present in the species responsible for bubonic plague. Perhaps the ability to cause rapid and severe clumping of blood and the subsequent clogging of capillaries accounts for the increased virulence of this one species.

This abstract was part of a written report submitted by Heather Campbell. Heather aroused your interest in the first paragraph by mentioning bubonic plague. She continued in the second paragraph to give a little of the background information that provided the idea for her project. The last two paragraphs include a summary of her methods, results, and conclusions.

STATE YOUR PURPOSE

The next section of your report is a brief introduction, which explains the general nature of your project. State what the project attempted to prove and what independent and dependent variables you tested. If you check your hypothesis, you'll probably discover that a simple rewording with a few additional remarks are all you need to write your introduction.

The introduction should also include relevant background information about your topic. Refer to the notes you wrote on index cards to refine your topic. Use any information that establishes the importance of your investigation. Review the significant literature in your own words, referring to a variety of sources (books, periodicals, pamphlets, encyclopedias, online references, etc.), and cite the references in a bibliography.

Be brief: your goal is to familiarize the reader with similar research work that laid the groundwork for your project. After discussing the purpose of your investigation, explain what impact your work might have on scientific knowledge or technological know-how.

EXPLAIN YOUR METHODS

Now that you have filled your reader with interest and curiosity about your project, explain how you carried out your investigation. You must carefully outline the exact process you followed so that anyone can repeat your experiment by simply reading the description in your report. You may want to include photographs or diagrams to illustrate complex points or sophisticated equipment.

Provide a complete list of equipment, materials, chemicals, specimens, and apparatus. Thoroughly describe all the steps and give complete instructions for building any equipment. Imagine what would happen if you bought an unassembled bicycle, only to discover that all the parts were included but that the assembly instructions were missing. Even if the instructions were provided, you could not assemble the bike correctly if they were incomplete or had the steps out of sequence.

Read the following description. Even if you do not understand the terms and abbreviation, you will be able to appreciate how specific Heather was in describing her procedure for growing and collecting bacterial cells.

> Bacteria were inoculated at 26°C for 24 hours. Organisms were washed in 0.033 M potassium phosphate buffer (pH 7.0) and inoculated into the defined liquid medium of Higuchi as modified by Zahorchak (Carter, P.B., et al., 1980, p. 638) at an optical density ($O.D._{260}$) of 0.1. Cells were incubated on a waterbath shaker at 26°C in Erlenmeyer flasks (10 ml/125 ml flask) until late log phase for two transfers. Final transfers were to the same media at an ($O.D._{620}$) of 0.1. Any pH adjustment was made at the final transfer. After incubation until late log phase, cells were spun down in the centrifuge (Sorvall RC2-B) at 10,000 rpm (12,000 g) for 20 minutes. The supernatant was poured off and bacterial cells were resuspended in phosphate buffer to a density of 4×10^{10} CFU/ml.

Heather's method for growing and collecting bacterial cells is detailed enough so that anyone could repeat her procedure from her report. She even provided a reference listed in her bibliography for anyone wishing to check the original paper for preparing the liquid medium.

This procedure provided Heather with the high concentrations of bacteria she needed in order to study how these organisms might be responsible for bubonic plague. You don't have to work with disease-causing bacteria, but every step of your procedure must be explained in as much detail as Heather's.

FOLLOW UP WITH RESULTS

The next section of your report should contain the results. In addition to a straightforward display of the data in rows and columns, use graphs and diagrams. Remember that your report should be comprehensive. Include any failures, errors, or results you can't explain. Don't analyze or interpret your data in this section; simply report the results in a clear and organized manner. You may have to spend time looking through your record book to group results from similar experiments.

CONCLUDE YOUR REPORT

After you have organized your results, the time has come for explaining, interpreting, and evaluating your data. Carefully examine your data to determine what they indicate. Is your hypothesis supported or rejected? Don't worry if your findings do not verify your hypothesis. Remember that failure to obtain supporting evidence does not mean the project is a failure.

Don't limit concluding remarks to an explanation of your results. Perhaps your findings have applications in other areas. For example, the results of a study of factors affecting the friction produced between certain materials may lead to improvements in car design. Be sure to mention any possible applications in your report.

Also, don't hesitate to recommend changes or improvements that can be made in your project. Any ideas to extend the project will show your interest in improving your investigative skills. To show how you can follow up on a clue, suggest possibilities for future experiments based on leads provided by your data. Be careful not to confuse conclusions with suggestions. Base any conclusion on analysis of your data, and any suggestion on extension of your data.

INCLUDE YOUR REFERENCES

The last section of your report should be a list of all your references in the form of a bibliography. A bibliography contains all the books, encyclopedias, pamphlets, and periodicals you used while researching your topic. Organize this list in alphabetical order.

The following forms are generally used in citing a reference: for a book—last name of author, first name, title, city of publication, publisher, date of publication; for a magazine or journal article—last name of author, first name, title of article, name of journal, volume number, pages. Accuracy is important, because you want to give proper credit to any scientist whose work provided the basis for your project. The bibliography at the end of this book will show you how to prepare your own bibliography and also help you locate other references that might help you with your project.

Acknowledge anyone who provided advice and suggestions in a section that follows your bibliography. Also include people who donated materials, allowed use of their facilities, or helped with the construction of equipment. As part of your written report, include any forms that have been signed by a teacher or professional verifying that you have followed the recommended guidelines in using hazardous materials, vertebrate animals, or human subjects.

SCIENTIFIC WRITING: CONTENT

Be sure your report answers all of the following questions.

- What problem or question is being investigated?
- What background information exists on this topic?
- What equipment and materials were used in your experiment?
- What procedures were followed to solve the problem or answer the question in this project?
- What observations were made during the course of the investigation?

- What information and data were recorded?
- What conclusions were made regarding the original problem or question?
- What suggestions were included for further research work to solve the problem more convincingly or answer the question more thoroughly?
- What new problems or questions were uncovered by the project?
- What sources were used?

SCIENTIFIC WRITING: STYLE

Make your report clear, simple, and accurate so that everybody can understand your work. Don't assume that someone reading your report is scientifically well versed; no one can be knowledgeable about all areas or disciplines of science. While some people may not appreciate the full significance of your investigation, your report should provide a clear understanding of its general nature and significance.

Clarity is the direct outcome of simplicity. Every word in a scientific paper should have a purpose. Don't use more words than necessary. You may think that the longer the report, the more impressed your readers will be. Consider, however, that the readers you must impress are the science fair judges. Impress them with quality, not quantity.

To improve the quality, begin by realizing that you're not finished after the report has been written—it's only the first draft. Go over your paper several times to see how words can be eliminated, sentences made simpler, or paragraphs clarified. Your written report is the *only* way everyone can appreciate what you've accomplished, so give it as much thought, enthusiasm, and effort as you devoted to conducting your project.

Don't be whimsical in presenting data. Information presented in a disorganized manner reflects uncertainty and may require more time for interpretation than your reader is willing to spend. Check for organization and continuity in your report. Be sure that the content within each paragraph is interrelated. Use transitional sentences to connect sections.

Be totally objective—don't exclude data that conflict with your position. As a scientist, you must not exhibit prejudice in drawing conclusions. Your report must clearly show the steps you took in analyzing your data to reach a conclusion.

The need for accuracy is self-explanatory. A scientific report implies content based on recorded observation and documented evidence. You want the reader to understand what you did *and* agree with your conclusions. Your report should convince your readers that your interpretation of the data is the only possible explanation.

To determine if your report is scientifically accurate, ask your science teacher to read it. To check its style, grammar, punctuation, and spelling, ask your English teacher to read it. After these teachers read your report, you may have to rewrite it. Rewriting may seem tedious, but you will discover that the result is worth the effort. If you use a word-processing program to prepare your report, rewriting will be not be too difficult.

If your project wins an award at the fair, you may want to submit it to a scientific journal for publication. Your science teacher may be able to suggest which journals are appropriate for your topic. Even if your report does not appear in a scientific journal, most fairs publish the abstracts of the winners. Your written report can also be featured in your school's newspaper or newsletter. Not only will your published report reflect your hard work and enthusiasm, but it will also serve as inspiration for others who plan to do a science fair project in the future.

CHAPTER 7

Graphing and Analyzing Your Data

Simply displaying your data may not be the most effective way to show what your project accomplished. Rather than placing your data in columns and rows, use a graph or chart to show a clearer relationship between the variables. There are three common types of graphs: line, bar, and pie.

Each type of graph highlights information differently. If you tested the effects of an independent variable, use a line graph. If you analyzed the relative effectiveness of several products or show the frequency of a particular observation, present your data with a bar graph. If you conducted a survey, display your results with a pie chart. You can display your data in all three ways and then decide which is best. By the way, there's no reason why you can't design an entirely different format if you're not satisfied with any one of these. You can also use a computer program that is designed to create and draw graphs.

LINE GRAPHS

Several different kinds of paper are available for making line graphs, each with a specific purpose. Select the one best suited to your needs. You may have to experiment with different types before deciding which to use. Your math teacher may be able to help you. No matter which kind of graph paper you choose, be sure that you follow the correct procedures for plotting data.

Always place the dependent variable on the vertical axis and the independent variable on the horizontal axis. Each axis should be clearly visible. Next, plot your points as accurately as possible. Be consistent when assigning your values for the vertical and horizontal axes.

Let's say you want to make a graph that shows how the solubility of solutions of lithium chloride (LiCl), sodium chloride (NaCl), and potassium chloride (KCl) changes with temperature. You decide to have each division on the vertical axis rep-

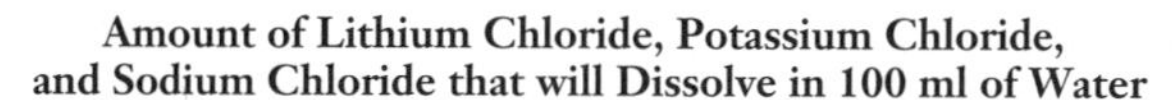

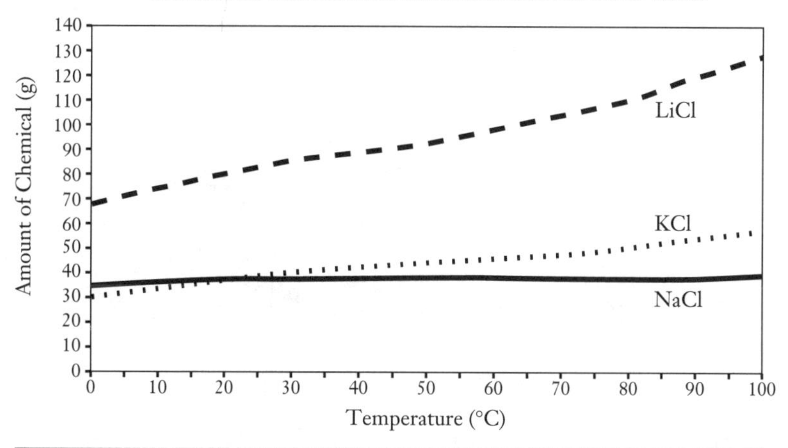

When graphing data, label the axes consistently. Clearly distinguish different experimental results with labels, symbols, or colors.

resent 10 grams, while each division on the horizontal axis represents 5°C. Once you have decided the value for each division, you cannot arbitrarily vary the scale. For example, you cannot change a division on the vertical axis to represent a 5-gram increase.

Draw the graph to fill the paper as much as possible, rather than squeezing all the points into a small section. If you spread out the data, it will be easier to see any slight, but important, difference between two points. When plotting the results of several experiments on the same graph, you can use symbols or different colors to distinguish between them. Be sure to include a key that shows what each symbol represents.

BAR GRAPHS

To construct a bar graph, draw a rectangular box from a point on the vertical axis down to the horizontal axis. As with a line graph, place the dependent variable along the vertical axis and

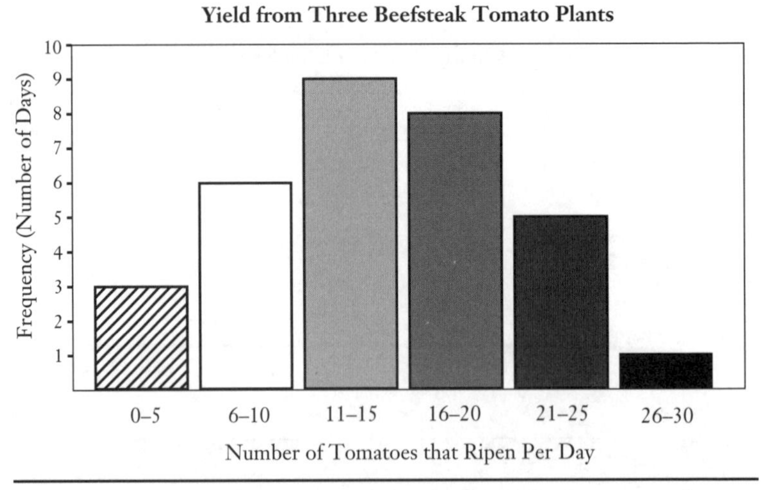

Clearly distinguish different experimental results drawn on the same bar graph. You can fill in bars with different patterns or use different colors.

the independent variable along the horizontal axis. If you want to show the number of days that have a particular yield of beefsteak tomatoes, you could create a bar graph like the one on the previous page.

Again, be consistent when constructing the scale along each axis. In cases where you record several experiments on one graph, distinguish the results for a particular investigation by filling each bar with a different color or a different pattern.

PIE CHARTS

A pie chart consists of a circle divided into sections. Pie charts are generally used to show the results of a survey. The whole pie may represent the total number of people interviewed, the

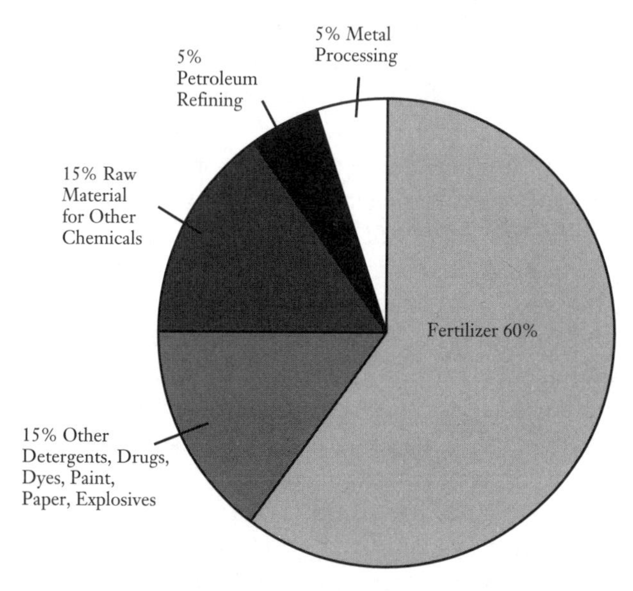

A pie chart can be useful when showing the results of a survey. Many computer programs can generate impressive-looking pie charts.

total amount of sulfuric acid used in industry, the total number of bacteria samples collected, or the total number of dogs living in your town. Each section of the pie stands for the proportional size of one of the groups in the survey. The size of each section is calculated by determining the percent for a particular category and multiplying that number by 360 degrees (the total for the circle).

For example, if 60 percent of the sulfuric acid used in industry becomes a component of fertilizer, multiply 0.60 by 360 degrees to obtain 216 degrees. Using a protractor, draw a section with an angle of 216 degrees. Divide the rest of the circle in a similar manner. Pool any results where the numbers are too small, and label this section "other."

ANALYZING YOUR RESULTS

Your results are significant only if you arrive at a specific conclusion. Spending a considerable amount of time in carrying out experiments only to find that your results are inconclusive can be quite disheartening.

For example, if you tested a chemical for its ability to stop cell division, you want to be sure your results are due to the action of the chemical and not to chance. A statistical analysis can indicate if your results are meaningful: Is there a significant difference between your experimental group (cells exposed to the drug) and control group (cells not exposed to the drug)?

Although such a statistical analysis may sound complicated, the calculations are quite simple. In fact, anytime you determine an average, you are performing a statistical analysis. An average or an arithmetic mean can help you decide whether a significant difference exists between the experimental and control groups.

The median is another statistically derived number, representing the middle measurement. There are equal numbers of measurements above and below the median. The median can help you get some insight into the distribution of results. Two different experiments may have the same average result, but the median can reveal some interesting differences.

Other statistical tests include determining the standard deviation, standard error of the mean, goodness-of-fit (chi-square test), and linear correlation. The calculations used for each of these procedures are beyond the scope of this book. However, you can see how to do these analyses by referring to a book such as *Means and Probabilities: Using Statistics in Science Projects* by Melanie Jacobs Kreiger.

If you work with computers, you may want to write a program that analyzes and graphs your data. Such a program can be an impressive part of your report. Whether you analyze by calculator or computer, just be sure you know enough about the simple kinds of statistical analyses, especially if you are working with a large amount of data. Only then can you reach a valid conclusion.

CHAPTER 8

Preparing Your Project for Display

When your project is finished, the time has come to prepare it for presentation at a science fair. You must prepare a display, summarizing what you have accomplished. Because you will not be able to include everything in your display, you will have to be selective. Be sure to include anything original, unusual, or creative. Select interesting highlights and impressive illustrations to attract the judges' attention.

BUILDING YOUR DISPLAY

Science fairs often feature such a large number of projects that regulations usually limit the dimensions of each display to 30 inches from front to back, 48 inches from side to side, and 48 inches higher than the table. Most displays follow a basic format that includes three, self-supporting panels. Your project can really stand out if

You may want to experiment with the format you use to exhibit your project. Rather than choosing a traditional three-sided display, Kathy Mitchell decided to display her project using a fan-folded arrangement.

you use a unique design, perhaps using a different geometric pattern for the panels, as Kathy Mitchell did with her project.

Whether you choose the traditional format or an original design, use sturdy materials such as plywood, masonite, pegboard, or heavy cardboard. The figure on the next page shows how you can obtain all the pieces needed to build two traditional three-sided displays from a 4- x 8-foot sheet of card-

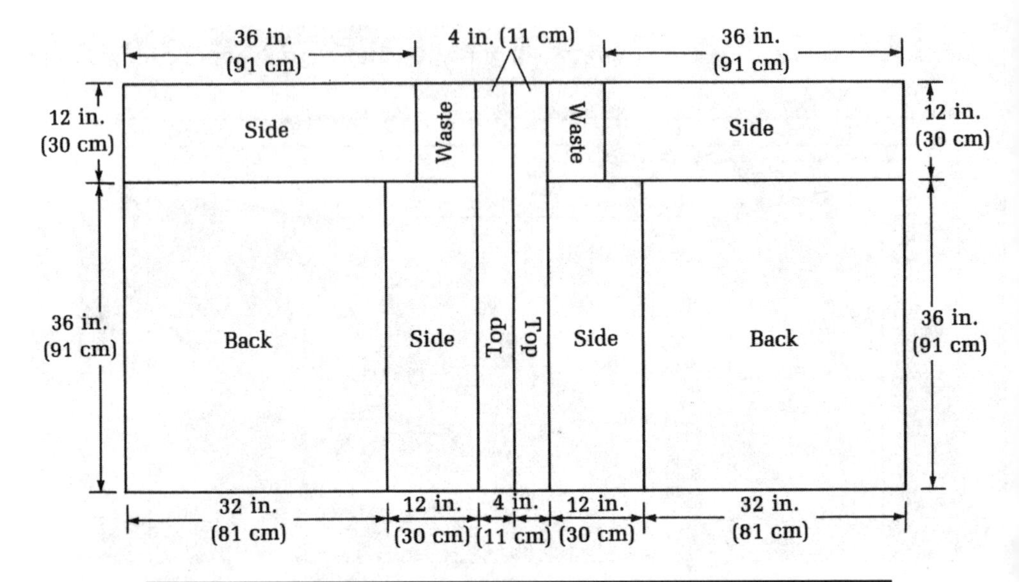

You can build two traditional three-sided displays from a single 4- x 8-foot piece of material. Cut the pieces as shown in this figure.

board or wood. Paint or cover the material with colored paper to avoid an unfinished or rough appearance. Setting up and breaking down the display will be much easier if you attach the three panels with hinges.

Before constructing the display, make a sketch drawn to scale to help you plan where to place all the items on the panels. Be as detailed as possible, indicating exact sizes, shapes, and positions. Assemble a mock-up out of paper or cardboard to help you decide where to position the items.

The lettering for any titles on your panels should be neat, plain, and readable from a distance of 2 meters (6.5 ft.). You can cut the letters from colored construction paper and attach them to the panels with rubber cement, glue, or tape. Apply as little adhesive as possible to avoid wrinkling the letters. Do not use staples, because they will be visible and detract from the appearance of your lettering.

You can also draw the titles with stencils and then color them, or use stick-on letters, known as Prestype, available from office-supply stores. Prestype comes in sheets that contain letters, numbers, and symbols. These are available in several different sizes. To transfer the Prestype to your display, place it over the paper or cardboard and then rub gently with a pencil or blunt-pointed object.

Carefully check the spelling in your titles and other parts of your display. Draw your graphs and charts neatly. Type your text material rather than writing it. If your display includes misspelled words, sloppy graphs, or illegible writing, the judges may have a negative opinion of your project before you've had a chance to talk about your work.

DISPLAY PHOTOGRAPHS AND DRAWINGS

If you have a camera, take photographs to record various stages of your project. These photographs can show any large equipment or research facilities used to conduct your investigation. You can also display photographs of your model at various stages of construction, people used in your survey, or any specimens too large to include in your display.

Use 5- x 7-inch or 8- x 10-inch black-and-white or color photographs; smaller pictures won't show important details. Color is preferable to black-and-white for attracting someone's eye. However, black-and-white photographs can be quite striking, provided the contrast is good. Don't staple or glue the photographs to your panels; mount them on cardboard to enhance their appearance. A salesperson at a local photo shop may be able to provide additional tips for taking and displaying photographs.

In preparing any drawings or charts, avoid materials that smudge, such as chalk, pastels, crayons, or charcoal. While handling your display, you may accidentally smear the illustrations. If you must use something that might smudge, cover it with clear acetate until the last possible moment.

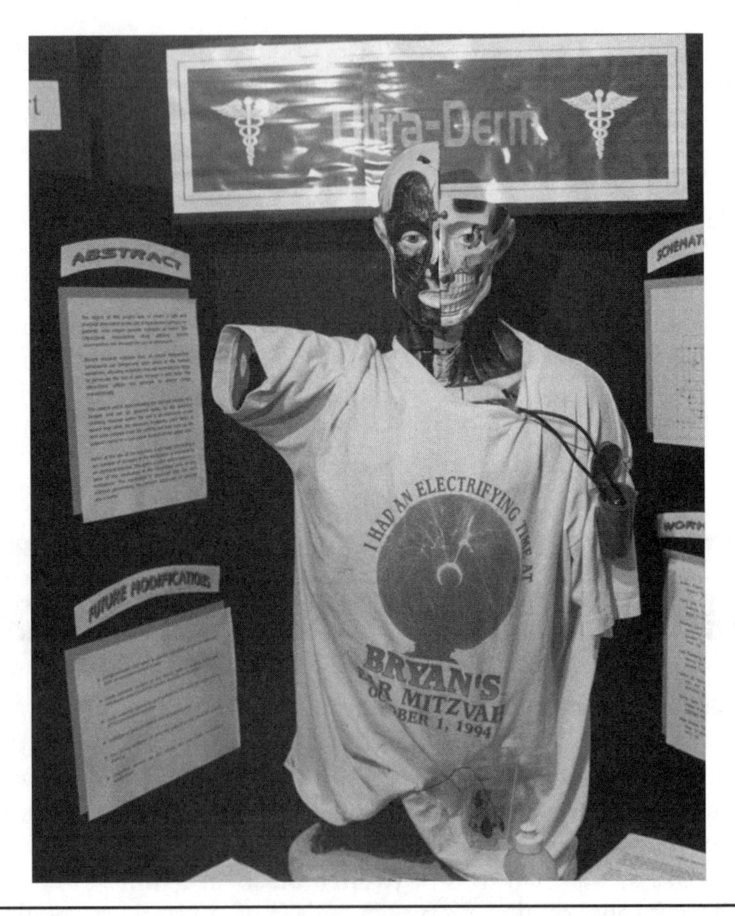

Avoid using large specimens or models in your display because they will detract from your overall efforts.

Don't mount your photographs and drawings in an arbitrary fashion. Try to organize them in an eye-catching arrangement. Select the most important visuals from your project. Some displays lose their appeal because too much is included and the highlights of the project get lost in the confusion. Also, too much color or gaudy color combinations will detract from the visual appeal of your display.

Be sure to arrange your display neatly and logically.

Because you want your display to be seen in the best possible light, bring one or two portable lamps from home. You may find that lighting conditions in the exhibition hall are poor, making your display difficult to see and impossible to read. If there is an electrical outlet near your table, use the lamps to illuminate your display. Make sure that the lights do not distract from your display by producing a bright glare.

WHAT ELSE TO DISPLAY

If you display a model, it should look neat. Be sure to sand the surfaces and polish metal parts. Use paint to make your model as attractive as possible. Make sure your model doesn't fall to the ground in all the confusion of setting up your exhibit. Attach it to a plywood base if necessary. You may want to include diagrams and photographs of the model as well as information you acquired as a result of constructing your model.

For a display project, exhibit samples of the most interesting and unusual specimens in your collection. If you cannot include them because they are too large or dangerous, use photographs. Also display some written material that explains the scientific knowledge you've gained from your project.

It is possible to show the equipment and materials used in your project without making a display that seems cluttered and disorganized.

A display of a survey project will probably contain more written material than one for a model or display project. But if you include only written material, your display may not look too appealing or interesting. Use some relevant illustrations, colorful graphs, or informative charts. Add any small touch that will improve the appearance of your display. Place your written information on a color background, enclose the questionnaire used in your survey in a plastic binder, or include some photographs of the people you contacted.

If your project involved lab work, place your complete report, materials used in the investigation, or samples from your experimental group on the table in front of the display. In that way, the backboard, side panels, and table comprise a comprehensive exhibit of your project.

WHAT TO EXCLUDE FROM YOUR DISPLAY

Exclude items that do not provide much information about your project. Displaying a large number of items, does not always make a project more comprehensive. Too many items may detract from the important points of your display. Start with materials that convey the most information. You can use the other materials if you have space left over.

You may not be allowed to exhibit anything considered hazardous to the public (disease-causing organisms, microbial cultures, food materials, dangerous chemicals, combustible gases, flames, gas tanks, glassware such as syringes and pipettes, or certain types of lasers).

Any displays involving an operating device may be subject to certain restrictions, including the prohibition of high-voltage equipment, batteries with open-top cells, bare wiring, and temperatures exceeding 100°C (212°F). All electrical wiring must be properly insulated, and all switches must be located out of the reach of observers. Science fair officials have the right to remove any display considered hazardous or potentially dangerous, so be sure to check the rules and regulations before your arrive at the fair.

There's another important thing to do before you take your display to the fair: prepare your oral presentation. This is your opportunity to explain what you have done, so consider your presentation as important as any other part of your project.

CHAPTER 9

Presenting Your Project

Don't wait until the day of the fair to begin preparing your talk for the science fair judges. Not only will you be busy making last-minute preparations on your display, but you will also find it difficult to think in the exhibit hall, which will probably be noisy and confusing. All the participants will be scampering about looking for their assigned spaces and hauling their projects into the hall.

PRACTICE MAKES PERFECT

Plan what you will say to the judges well in advance of the fair. Rehearsing your talk with your parents, teachers, and friends is an excellent way to practice your presentation. Listen to a tape recording of your presentation to hear how it sounds to others. If you are not satisfied, make changes and then repeat the process. The more you practice, even in front of a mirror, the greater your chances of impressing your listeners. You may have as little as 5 minutes—and certainly no more than 10—to present your project, so effective communication is necessary.

Because eye contact is important in keeping the judges' attention, do not read your presentation from note cards. You can refer to an outline to guide you through the presentation, but don't read the report word-for-word or repeat it from memory. If you do, you are likely to deliver your presentation in a monotone voice. The judges might find your presentation boring. In addition, you're more likely to get flustered if a judge interrupts you to ask a question. Deliver your speech with spontaneity and enthusiasm. Project your voice so that all the judges can hear.

HOW TO LOOK AND ACT

Because judges will be looking at you during the presentation, don't chew gum, wear old or dirty clothes, sway back and forth, or peer into the distance. People are impressed with good manners, appropriate dress, polite behavior, and an interested attitude. Even though your behavior has nothing to do with the merits of your project, the judges might get a negative impression from your appearance that carries over into their evaluation of your project. Just think about a time when you may have formed a negative opinion of someone solely on the basis of his or her appearance.

Be sure not to block the judges' view by standing in front of your project. Stand to one side so that the judges can clearly see your exhibit. Get them actively involved by giving them a copy of your abstract or arouse their interest by handing them a live specimen. To maintain their interest, make periodic eye contact with each of the judges. While speaking, be sure to point out any laboratory apparatus, charts, or photographs on display to reinforce your points.

When presenting your project, be sure to stand to the side so that the judges can clearly see your display.

WHAT TO SAY

Begin by introducing yourself. Then give the title of your project, followed by an explanation of your purpose. Briefly summarize any background information and discuss how you developed an interest in the topic. Explain how you proceeded with your project and emphasize any results or conclusions by pointing to graphs, charts, or tables. Be sure to mention any problems encountered in completing your project, and tell the judges how you overcame these obstacles.

If you did a laboratory investigation, go directly from your results to your conclusions. Again keep in mind that failure to support your hypothesis is perfectly acceptable. Tell the judges what applications your project has to other areas of science or point out its relevance to everyday situations. Comment on any ideas or suggestions for further research or ways to improve your project.

Finally, invite questions from the judges. Think before you answer. Speak slowly when talking to the judges. If you are asked a question you feel is not related to your project, tell them that you were not concerned with that particular area in your work. However, don't hesitate to say that you don't know the answer to a pertinent question. You can express your opinion or venture a guess, but admit to the judges that you are only giving your opinion or guessing. Any attempt to talk around a judge's question might be disastrous.

Be direct and honest in acknowledging any help and assistance you obtained in conducting your project. As professionals working in science and engineering, the judges will be able to determine how well *you* understand the nature and implications of your work. When you have finished your oral presentation, thank the judges for their time and interest.

WHAT THE JUDGES LOOK FOR

The judges are not there to criticize your work. They realize you may have spent considerable time and effort to complete the project, so they are interested in what you have learned from it. They are especially looking for ways in which you have approached and conducted your project in a scientific manner.

Ideally, the judges should be more interested in promoting and fostering your interest in science than in evaluating your project for an award. As in all contests, however, the judges must choose winners and award prizes. Not everyone at the science fair may feel they received the recognition they deserved. You may be more understanding if you recognize some limitations imposed on judging at a science fair.

Consider the how little time you have to impress the judges. You may have conducted your project over a period of several years, but you have no more than 10 minutes to convey its importance to the judges. In addition, consider that only one person may judge your project at a local or regional fair. In larger competitions, a team of several judges may be used. These people are volunteers recruited from schools, universities, research centers, industries, hospitals, and government agencies.

When judging teams are organized, an attempt is made to match their background and expertise with the nature of the project. A doctor *should* judge a project investigating the effects of antibiotics on disease-causing organisms. An engineer *should* evaluate a project exploring new designs for airplane wings.

However, the selection of judging teams may be a matter of chance. The doctor may wind up judging the wing designs, while the engineer may question how the antibiotic was tested. As a result, an excellent project may not get recognized for its sophistication of design and quality of work.

Keep in mind that judging is subjective, even though the judges have guidelines to follow. Judges may have personal preferences or pre-formed opinions. After all, judges are human. In any case, they may fail to give the proper recognition to a project. At the same time, most projects probably don't win prizes simply because "better" projects were around. The main thing, of course, is to do your best and not worry about trying to second-guess the judges.

JUDGING CRITERIA

Not all science fairs use a standardized scoring sheet for judging projects, nor do they adhere to the same point distribution for each category. Nonetheless, most judges evaluate the project in terms of five basic areas: (1) scientific content and application, (2) creativity and originality, (3) thoroughness, (4) skill, and (5) clarity. Within each area, a number of questions can be asked. The judges may rate each answer on a scale from 0 to 5, with 5 being the highest rating.

Try to balance what the judges expect with what you can do within the limitations of your time, resources, and ability. In addition, do not conduct a project solely to impress the judges. The project must reflect your interests and the need to please yourself. In that way, a natural enthusiasm for, and genuine pride in, your work will be evident.

SCIENTIFIC CONTENT AND APPLICATION

The judges will check to see if you approached your project scientifically. Did you state your hypothesis clearly? Was the question or problem sufficiently limited so that an answer or solution was possible? Hopefully, you narrowed the topic so that it could be investigated and explored in a reasonable amount of time. Simply working on a difficult and complex problem without arriving at a solution is not impressive.

A good scientist can identify an intriguing problem that is solvable within a reasonable amount of time using the most technologically advanced equipment. In the case of a scientist, that time may require several years of concentrated work. Science fair judges will realize that your project must reflect your age, involvement in school and community activities, and available resources.

The judges will also question you on procedures and methods. Did you design and execute your plan efficiently? Did you clearly recognize and define all the variables? If controls were necessary, were they included and used correctly? Did you record sufficient data to support your conclusions? Were your data relevant to answering the question posed by the project?

Did you recognize the limitations of the data and see connections between the project and other areas? Did you include ideas or specific proposals for further research? Was your bibliography comprehensive, and did it cite scientific rather than consumer publications?

CREATIVITY AND ORIGINALITY

The judges may ask how the project originated—did you come across the idea while reading a textbook, talking to a scientist, or going through literature? If the idea was the independent outcome of your work and effort, the judges will give appropriate credit for creativity.

While no penalty is imposed if you started your project by seeking the advice of a professional or reading literature, no points for creativity are awarded to an impressive project copied from a textbook or developed entirely by someone else. A less sophisticated project that is genuinely your brainchild and work will get more credit for creativity.

Your project can receive credit for originality in several ways. Did you design your own procedures to perform some lab analysis? Perhaps you constructed a piece of equipment specifically for your project. Of course, you may need to construct a kit to carry out your project. You might purchase a kit to build a telescope that will allow you to observe Halley's comet. In these circumstances, the judges would not penalize the project, because they recognize that the telescope is only a tool, and not the product of the project.

Any student can spend money to buy equipment, but creative students will devise their own. Such students are always coming up with new ways of answering old questions. By the way, if your project involves some field of engineering, the judges may not give high marks for creativity to a device that is ingenious but inoperable.

Originality can also be displayed in your analysis of data. Was there more than one way to interpret the data? Do conflicting results point to some new directions? Are additional experiments needed to answer problems raised by inconclusive data? Don't extend yourself too far, but use your imagination to suggest further experiments or discuss the application of your project to other areas.

THOROUGHNESS

The maximum number of possible points for thoroughness is usually not as high as the number awarded for scientific content and creativity. The judges might ask whether you based your conclusions on a single experiment or on enough repetitions to obtain sufficient data. They may examine your notebook to determine if you kept complete and accurate records for each experiment.

The judges may also question whether you used all possible approaches to test your hypothesis. You may be asked to comment on other theories concerning your project or additional ways of interpreting your data. They may also look through your bibliography to see if you were thorough in your library research.

SKILL

Judges usually consider technical skill to be as important as thoroughness. They may ask questions to determine if you have all the design, laboratory, observational, analytical, and construction skills necessary to complete your project. Naturally, the judges will be interested in the amount of assistance provided by parents, teachers, scientists, and engineers. They may ask about the equipment used in your project: Was it built independently or was it located in a laboratory where you worked?

CLARITY

Finally, the judges will evaluate the clarity of your presentation and exhibit. How well did you explain and discuss your project? Obviously, this will reflect the extent to which you under-

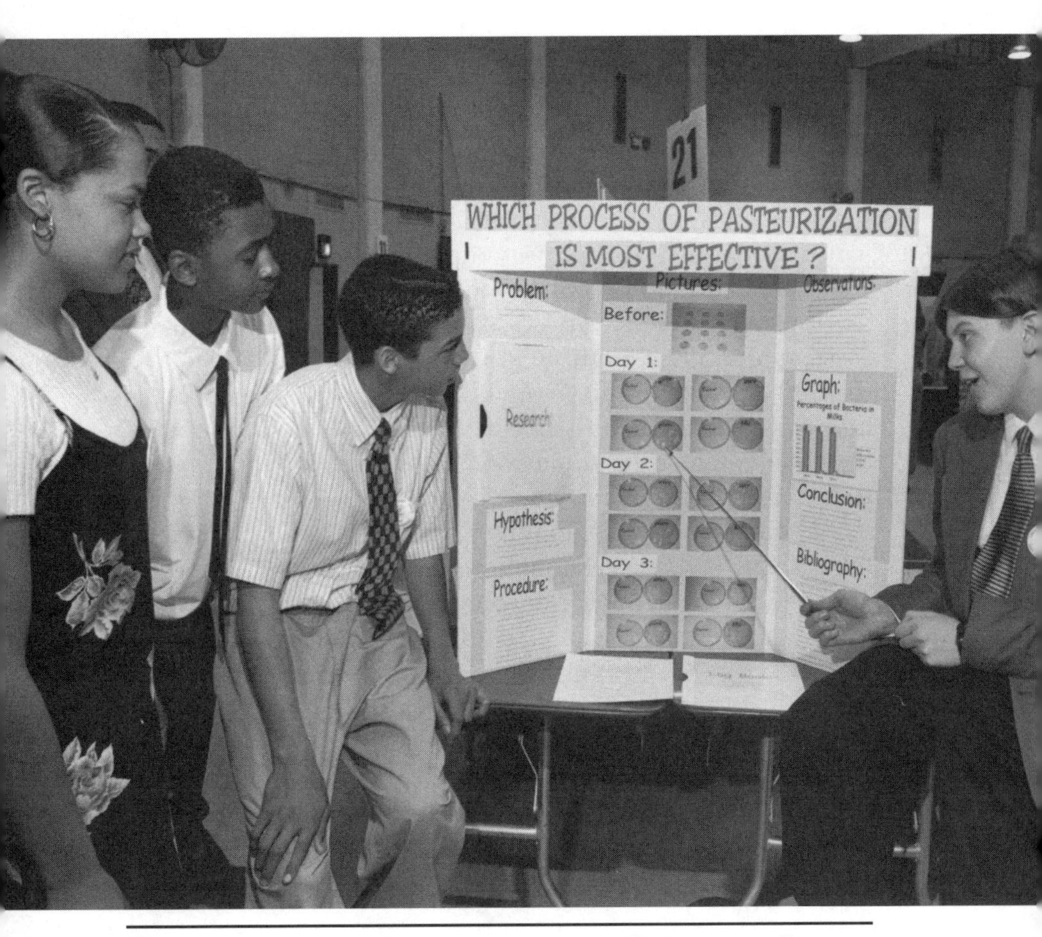

If you enter a regional or state competition, take time to walk around and talk with other students about their projects.

stand your investigation and its applications. Clearly explain your methods so that everyone can understand what you accomplished.

The judges will examine your exhibit. Were the important phases of the project presented logically and clearly, without resorting to flashy gimmicks or cute gadgets that detract more than they enhance? Remember, your project is not being judged for its special effects but rather for its scientific value.

MISSION ACCOMPLISHED

After making your presentation, the most anxious time has arrived—waiting for the announcement of winners. Winning is nice, but it's not everything. If you do not win any prizes at the fair, think about what you have accomplished. You began with an idea, planned a project, explored the world of science, and built an exhibit to display your own work.

If you don't receive an award, listen carefully to the judges' suggestions. They might help you improve your project for the next science fair. But more importantly, take the opportunity to walk around and talk with other students. See what they did; you may get some good ideas for next year's project from this year's winners.

The experience of completing a project and exhibiting it for evaluation by professionals is worth the effort, no matter what the outcome. Approach your project as a learning experience, not as a way of winning a prize at a science fair.

GLOSSARY

abstract—a short, detailed, written description of a scientific research project.

control—part of an experimental setup permitting only one independent variable to be present.

control group—a test group used as a basis for comparison where no experimental factors are introduced.

dependent variable—a factor in the experiment that is caused to change or is affected by a second factor under the experimenter's control.

experimental group—a group being subjected to the factor being tested in the investigation.

hypothesis—a statement or idea to be proved or disproved by experimental testing.

independent variable—the factor that the experimenter can change at will.

inference—a conclusion that is not based on direct evidence or observations.

invertebrates—animals without backbones, such as crabs, clams, caterpillars, and cockroaches.

qualitative experiment—a procedure whereby observations, but not numerical results, are recorded.

quantitative experiment—a procedure whereby measurements and numerical data are recorded.

statistics—a branch of mathematics dealing with the analysis of numbers.

vertebrates—animals with backbones such as fish, frogs, finches, and humans.

BIBLIOGRAPHY

BOOKS

Ardley, Neil. *101 Great Science Experiments*. New York: Dorling Kindersley, 1993.

Barr, George. *Science Research Experiments for Young People*. New York: Dover Publications, 1989.

Bochinski, Julianne Blair. *The Complete Handbook of Science Fair Projects*. New York: Wiley, 1996.

Bombaugh, Ruth. *Science Fair Success*. Springfield, NJ: Enslow Publishers, 1990.

Bonnet, Robert L. *Science Fair Projects: The Environment*. New York: Sterling Publishing Co., 1995.

Bonnet, Robert L. and Daniel Keen. *Botany: 49 More Science Fair Projects*. New York: TAB Books, 1991.

Bonnet, Robert L. and Daniel Keen. *Computers: 49 More Science Fair Projects*. New York: TAB Books, 1990.

Collins, Kevin and Betty Collins. *Experimenting with Science Photography*. New York: Franklin Watts, 1994.

Dashefsky, H. Steven. *Environmental Science: High School Science Fair Experiments*. New York: TAB Books, 1994.

Durant, Penny Raite. *Prize-Winning Science Fair Projects*. New York: Scholastic, 1991.

Gardner, Robert. *Robert Gardner's Favorite Science Experiments*. New York: Franklin Watts, 1992.

Gardner, Robert. *Robert Gardner's Challenging Science Experiments*. New York: Franklin Watts, 1993.

Iritz, Maxine Haren. *Winning the Grand Award: Successful Strategies for the International Science and Engineering Fair*. New York: TAB Books, 1992.

Katz, Phyllis and Janet Frekko. *Great Science Fair Projects.* New York: Franklin Watts, 1992.

Krieger, Melanie Jacobs. *How to Excel in Science Competitions.* New York: Franklin Watts, 1991.

Krieger, Melanie Jacobs. *Means and Probabilities: Using Statistics in Science Projects.* New York: Franklin Watts, 1996.

Markle, Sandra. *The Young Scientist's Guide to Successful Science Projects.* New York: Lothrop, Lee and Shepard, 1990.

Newton, David E. *Making and Using Scientific Equipment.* New York: Franklin Watts, 1993.

O'Neil, Karen E. *Health and Medicine Projects for Young Scientists.* New York: Franklin Watts, 1993.

Provenzo, Eugene F. and Asterie Baker Provenzo. *47 Easy-To-Do Classic Science Experiments.* New York: Dover Publications, 1989.

Rybolt, Thomas. *Environmental Experiments about Renewable Energy.* Springfield, NJ: Enslow Publishers, 1994.

SERIES

Franklin Watts (New York). *Science Experiments and Projects* (eleven books). 1988–present.

Franklin Watts (New York). *Projects for Young Scientists* (twelve books). 1986–present.

Gardner, Robert. *Science Projects Series* (six books). Springfield, NJ: Enslow Publishers, 1994.

INDEX

ABOUT THE AUTHOR

Salvatore Tocci teaches chemistry and biology at East Hampton High School on Long Island, New York. His strong belief that science is best learned when students are actively involved has led him to write this book. In addition, Mr. Tocci has organized science fairs at his high school and judged both regional and national contests. He has also presented workshops for science teachers at meetings held throughout the United States. Besides *How to Do a Science Fair Project*, Mr. Tocci has written several books, including a high school text that focuses on the practical applications of chemistry.